The New Time and Space

Also by John Potts

AFTER THE EVENT: New Perspectives on Art History (*co-edited with Charles Merewether*)

A HISTORY OF CHARISMA

CULTURE AND TECHNOLOGY (*with Andrew Murphie*)

RADIO IN AUSTRALIA

TECHNOLOGIES OF MAGIC (*co-edited with Edward Scheer*)

THE FUTURE OF WRITING (*editor*)

THE UNACCEPTABLE (*co-edited with John Scannell*)

The New Time and Space

John Potts
Macquarie University, Australia

palgrave
macmillan

First published 2015 by
PALGRAVE MACMILLAN

Palgrave Macmillan in the UK is an imprint of Macmillan Publishers Limited, registered in England, company number 785998, of Houndmills, Basingstoke, Hampshire RG21 6XS.

Palgrave Macmillan in the US is a division of St Martin's Press LLC, 175 Fifth Avenue, New York, NY 10010.

Palgrave Macmillan is the global academic imprint of the above companies and has companies and representatives throughout the world.

Palgrave® and Macmillan® are registered trademarks in the United States, the United Kingdom, Europe and other countries.

ISBN 978–1–137–49437–5

This book is printed on paper suitable for recycling and made from fully managed and sustained forest sources. Logging, pulping and manufacturing processes are expected to conform to the environmental regulations of the country of origin.

A catalogue record for this book is available from the British Library.

Library of Congress Cataloging-in-Publication Data
Potts, John, 1959–
The new time and space / John Potts, Macquarie University, Australia.
pages cm
Summary: "The New Time and Space examines the ways in which the experience of time and space has been redefined in the contemporary era. Mobile phones, networked communications and freely accessible digital information have contributed to a shift in the way we understand time and space. There is a new virtual dimension to the public sphere, a new immaterial venue for social life. The book illustrates how, in this changed relationship to space, the virtual is layered onto the actual. Smartphone users are overwhelmed by a compulsion to check their messages, so that time becomes a montage of fractures. Interruption and distraction are the norm, as time is absorbed by an endless stream of messages and information" — Provided by publisher.
Includes bibliographical references and index.
ISBN 978–1–137–49437–5 (hardback)
1. Space and time in mass media. 2. Space and time—Psychological aspects. 3. Space perception. 4. Space—Philosophy. 5. Time perception. 6. Time—Philosophy. 7. Digital media—Influence. I. Title.
P96.S7P68 2015
304.2′3—dc23 2015019257

Contents

Figures

Acknowledgements

An earlier version of Chapter 4 appeared under the title 'The Theme of Displacement in Contemporary Art' in the journal *E-rea*, Vol. 9 No. 2. DOI 10.400/Erea.2475. © 2012 E-rea.

A portion of Chapter 5 was published in an earlier version as 'Journeys Through the Past: Contempt, Nostalgia, Enigma' in *Media and Nostalgia: Yearning for the Past, Present and Future*, edited by Katharina Niemeyer, Palgrave Macmillan, 2014.

The author wishes to thank the artists, their galleries and Kaldor Public Art Projects for their permission to reproduce artworks in this book.

Special thanks for encouragement and inspiration: Sophie and Leo.

Introduction

This book examines the ways in which the conceptualisation and experience of time and space have been redefined in the contemporary era. Mobile phones, networked communications and freely accessible digital information have contributed to a shift in the way we experience and conceptualise time and space. Globalisation and international network coverage have collapsed distance and delay in communication. Mobile networked communication fosters a form of virtual time and space, which is superimposed onto territorial space. Time is increasingly composed of interruptions and distractions, as smartphone users are overwhelmed by messages.

In the networked age, we are living with changed parameters of time and space. When the local is instantly and effortlessly linked to the global, our sense of space and distance alters. The scale of distance is overcome; telematic communication – online or with wireless phones – is co-present at a distance. There is a new virtual dimension to the public sphere, a new immaterial venue for social life. In this changed relationship to space, the virtual is layered onto the actual. The physical location – in geographical space – is less important than the virtual space – on the network. This tension between the two types of space is evident every day, when commuters are annoyed by a fellow user of civil space loudly enjoying a conversation on his mobile phone; or when parents try in vain to wrest their children's focus to the dinner table discourse and away from the texting or social media conversations currently absorbing their energies.

The New Time and Space explores the significance of these shifts within contemporary culture, while providing an intellectual history of time and space. In pursuing this theme, the book confronts the issues arising from the impact of digital technology and the internet. What has happened to the space of privacy in the age of social media and reality TV? What has become of the public sphere and public good in the era of personalised social media? Has online media made us better informed, or simply overwhelmed? In an age of smart machines and convergent media, why do individuals feel so short of time?

One feature of the book is its use of contemporary art. There are many examples cited from contemporary art, film, media and literature, on the premise that artists serve as the antennae of their society, picking up conceptual shifts and investigating them in creative works. The new time and space has been sensed, often in advance, by artists, and explored in many forms, including video art and performance.

It may be objected that my approach in *The New Time and Space*, focused on the effects of networking and digital media, is technological determinist. I have dealt with the question of technological determinism elsewhere;[1] it may suffice here to state that my focus in this book is on the cultural effects that ensue from the use of the new technologies. The properties of these technologies – digital mobile devices and global networked communications – are important in the possibilities they create. My approach in this regard has more in common with the German tradition of media theory, variously described as 'hardware theory' and 'media archaeology', than it has in common with much of the media theory and sociology of technology in the Anglo-American tradition. The German school – including the writings of Friedrich Kittler, Wolfgang Ernst and Bernhard Siegert – has pursued an investigation of 'the agency of the machine', in Ernst's phrase.[2] The dominant Anglo-American strand, by contrast, has been pre-occupied with the 'social shaping of technology' as traced in the political economy of media and technology; yet this social focus has often under valued the properties of the machines themselves. This shortcoming has become so evident in the age of the internet that the clumsy term 'affordances' has been imported from psychology, to describe the latent potential of objects.

Yet a substantial body of theory in media and technology studies exists – in English as well as in German, French and other languages – that has not needed recourse to 'affordances' to describe the role played by technology in cultural expression. This strand of thought can be traced back to Nietzsche, who wrote in 1882: 'Our writing instruments contribute to our thoughts'.[3] This aphorism is prescient of profound developments in contemporary culture. Nietzsche was writing – or typing – about the typewriter, but his idea has been tested – and contested – in the context of succeeding generations of mechanical and electronic technologies. That the properties of a medium can contribute at least in part to the nature of the transmitted message is now a familiar idea. That the technology of that medium affects the cognitive functions of those who use it is a more radical re-voicing of the same idea, and one consequently more likely to be resisted.

But the concept is explicit in Nietzsche's statement; it has been pursued in far greater detail by several theorists since the second half of the twentieth century. It has been expressed most famously in the work of Marshall McLuhan; more recently, writers such as Jack Goody, Walter J. Ong, Elizabeth Eisenstein, Joshua Meyrowitz and Pierre Levy have elaborated this notion in a more scholarly fashion. Goody developed the concept of 'intellectual technologies', asserting that writing creates a 'different cognitive potentiality'.[4] Ong, in his studies of orality and literacy, argued that 'more than any other single invention, writing has transformed human consciousness'.[5] Eisenstein elaborated a detailed account of the cultural and political effects of the printing press as 'agent of change'.[6] Meyrowitz, in *No Sense of Place*, examined the cultural impact proceeding from the specific technological form of television.[7] Levy, writing in the age of the internet, described hypertext as an 'intellectual technology' which modifies the 'cognitive ecology' into which it is introduced.[8] The approach pursued in all these works, in common with the texts of those German writers cited above, emphasises the properties of specific media and their effects in shaping sensory and cognitive activity.

My study of the new time and space, in the context of mobile networked technology, takes place within this theoretical tradition. But my focus is not primarily on technology: it is on the shifting conception and experience of time and space, culminating in the present condition. This is characterised as a heavily mediated hybrid social space, and a pressured sense of time experienced as a montage of fractures. Certainly social and political factors are considered in the formation of this condition, and the wilful technological optimism emanating from Silicon Valley – a re-voicing of an earlier, twentieth century, technocratic enthusiasm – is heavily criticised in this book.

I bring a historical perspective to developments wherever possible, tracing the development over time of ideas, practices and beliefs concerning time and space. One of these is the changing status of privacy, a form of space around the individual that came to be considered a right only recently; the public devaluing of this 'right' in the years since the advent of social media in 2004 is placed in a longer historical context. There was once little or no privacy; at one point it was constituted as desirable, then as essential, for civil society; within the 'radical transparency' of twenty-first-century social media, privacy has been decried as an obstacle to connectivity, rather than a right. The radically altered status of privacy across a relatively short span of history is a narrative comprising many elements: political, economic, social, cultural as well as technological. The properties of the technologies, and the possibilities they create, are one focus of my study in this book. But the new time and space is ultimately forged by the uses made – by individuals and by collectives – of the technologies of the twenty-first century.

Chapter 1 provides background for the contemporary conceptualisation of time and space. This chapter takes the very long view, drawing on ancient beliefs and the slow formation of concepts of time and space. It includes a survey of the varying means of understanding time and space through human history, from the mythological time and space of the earliest societies, to the philosophical inquiry of Ancient Greece, and the linear time conceived within the Judaeo-Christian tradition. The later development of scientific measurement and analysis is considered within the context of the empirical measuring and control of space and time. The impact of technologies such as the map, clock and calendar is weighed, as well as the technologies of communication and transport of modernity. The later electronic and digital technologies of communication are placed within an historical process beginning with the telegraph in the 1830s, whereby the message for the first time outpaced the messenger. The re-conceptualisation of 'space-time' generated by relativity theory and quantum physics is also discussed as a re-shaping of thought in the twentieth century.

Chapter 2 frames the analysis of contemporary time and space through a survey of the most influential theorising in recent years.

Included in the discussion are influential twentieth-century theorists of both time and space, as well as contemporary theoretical debates. This chapter includes discussion of the transformations of time and space perception in modernity, with emphasis given to Bergson's influential theory of duration and time. E. P. Thompson's assertion that each mode of production implies a different form of temporality is discussed, with reference to the 'time-discipline' of industrial capitalism. Henri Lefebvre's similarly Marxist perspective on the 'production of space' is also detailed. Alternative twentieth-century perspectives on space are voiced in Carl Schmitt's notion of *nomos* as orientation to space, and Bachelard's poetics of space. Contributions from Jameson, Virilio and Harvey to theories of postmodern space-time are considered. Castells' account of network society's 'timeless time' and 'space of flows' is assessed, along with Giddens and others on networked space and time. Recent findings in neuroscience concerning the perception of time and space are canvassed.

Chapter 3 contains the book's analysis of the contemporary experience of time and space. This chapter considers the ways in which our experience of time and space has been transformed by new communications technologies and networking. It proposes that a virtual network of time and space has been superimposed over the physical world, so that 'no one is where they are', as one respondent complained to psychologist Sherry Turkle in her recent book *Alone Together*. Individuals connected to the network via phone or laptop may be physically present but are in effect absent, engaged in a virtual conversation. Social media has created a new form of virtual space, in which individual users are frequently engaged. The physical world is increasingly overlaid by the virtual networked sphere: the term 'Hertzian space' has been used to describe the space of frequencies enabling connection. Mobile phones are also GPS devices, precisely locating users in space and enabling connection with others. Social media apps have been designed to 'read' social spaces, letting the user know how to interpret the people occupying that space.

Our time is more filled than ever with messages, constant interruption and distraction; even social media devotees frequently complain of having too little time. Much of this new communication is concerned with the present moment and 'phatic' communication for its own sake, resulting in an intensification of the moment and a loss of reflection and sense of the past. When so much activity occurs online, the notion of public space is transformed. When private phone calls are loudly enacted in public, the divisions between public and private have shifted. Our sense of time has been altered now that our days and nights are so filled with messages and information.

Chapter 4 discusses images and ideas of spatial displacement in recent works of art. The theme of displacement is considered in the context of the globalist aspect of contemporary art, itself a reflection of globalisation. The intensified movement of goods, information, capital, images – and people – around the world provides the setting for contemporary artists' treatment of displacement in a wide range of contexts. The chapter is illustrated with recent artworks from Documenta, the world's premier international art event, and elsewhere.

Chapter 5 discusses the temporality – or range of temporalities – found within contemporary consumer network society. One concern of this chapter is how contemporary culture, so enamoured of the new and the present, is informed by a version of technological progress, particularly in the context of information technology. The world of Web 2.0 and social media is a world fixated on the present and the near-future; the past is already gone and is barely relevant. The construction of the past in commodity culture is appraised as one of contempt: the present evinces an attitude of contempt for the inferior and 'primitive' condition of earlier times. In this regard, nostalgia is construed as the 'ideological twin' of progress, in that the past is represented as other, backward and ultimately undesirable. I discuss alternative perspectives on the past found in the work of contemporary artists, where the past is represented not as nostalgia or commodity, but as part of an ongoing dialogue with the present. In the work of Anselm Kiefer, Thomas Demand, Walid Raad and William Kentridge, the past has an enigmatic quality, devoid of the sentimentalising aspects of nostalgia, existing in the present in a complex, often unsettling, manner.

Chapter 6 considers the profound impact of various technologies on privacy, which is interpreted as a form of space. Privacy is the right to one's personal space. It is the right to control the immediate physical space, often in the domestic environment, as well as the right to control the immaterial sphere of information on the self. The use of networked information for 'data-mining' is a recent violation of that informational space. This chapter provides a brief history of the idea of privacy, linked to the effects of surveillance technology and networked culture. The profusion of cameras and telephoto lenses, the obsession with celebrity and the explosion of social networking have all intruded on the space of privacy, so that the status of privacy has been fundamentally changed by the growth and acceptance of these technologies.

Chapter 7 examines the political and social consequences of the networked distribution of digital photographs. Consideration is given to the power of social media to affect political discourse through the distribution of photographs, with reference to the role of blogs and photographs in the Arab Spring. The role of photographs on Instagram, Facebook and other social media is also discussed, as part of the construction of the 'relational self'.

Chapter 8 investigates the properties of sound in digital technology. Software programs enabling the manipulation and editing of music and sound render audio in visual terms as a waveform. This visualisation of sound through digital technology is an instance of the spatialisation of time which Henri Bergson criticised as a misunderstanding of the nature of time. This chapter explores what it means to 'see' time in the form of visualised sound, whether it amounts to a reduction of the richness of time, or makes possible creative play with the properties of time and sound.

In Chapter 9, works of recent video and performance art are appraised as investigations of the current experience of time and space. Many of these works, by artists such as Nam June Paik, Bill Viola, Shaun Gladwell and Daniel Crooks, are conceptual explorations of the way space and time are perceived and experienced: hence they represent an intellectual construction, as well as creative imagining, of 'capsules' of space-time. Christian Marclay's *The Clock* is analysed as an extended meditation on mechanical time and the experience of duration. Performance art works, particularly the extended 'ordeal' works of Marina Abramovic and Mike Parr, are treated as engagements with time and the multi-layered character of duration. This final chapter also includes a brief discussion of recent literature and film, in which the complexities of contemporary time and space have been creatively assayed.

1
A Brief History of Time and Space

Before time and space

In the beginning, there was no time or space; in their absence, there was void or chaos. The cosmologies constructed by the human mind – whether expressed as mythology, religion or contemporary physics – have posited a state of formlessness, or of nothing, prior to the existence of time and space. Albert Einstein, whose theory of relativity occasioned a re-conceptualisation of time and space in the early twentieth century, noted that 'scientific thought is a development of prescientific thought. As the concept of space was already fundamental in the latter, we should begin with the concept of space in prescientific thought'.[1] This insight pertains as well to prescientific concepts of time and to mythological accounts of the origin of time and space.

For long sections of its history, humanity has understood time and space through the prism of mythology, which provided narrative descriptions of the origin and nature of the world. Religion provided a later account, as did philosophy and, most recently, science. The experience and conceptualisation of time and space have been rendered in many forms, according to many systems of thought; there are striking differences, as well as surprising similarities, between the earliest mythological descriptions of the origin of the universe and the Big Bang theory of contemporary physics.

Cosmogony – the theory of the origin of the universe – is first articulated in mythology. These narrative accounts of the beginning of the universe are also descriptions of the birth of time and space. The mythologies of traditional, pre-literate peoples often proposed an initial state before time and space, characterised by darkness, lack of form or, frequently, water. 'In the beginning there was no land, there was but one water': this phrase from the origin myth of the Kets typifies, according to Murad Akhundov, 'the overwhelming majority of cosmogonic myths', including the Sumerian Nammu, the Egyptian Nun, the Indian Asat and the Babylonian Apsu.[2] Marie-Louise Von Franz, in her survey of creation myths, cites a North American example: 'In the beginning there was no Sun and no Moon and there were no stars. Everywhere there was darkness and water.'[3]

Water recurs as the description of a state before the formation of the universe due to its amorphous nature: it represents the formless, indefinite disorder out of which form and order are created. Mircea Eliade cites a Polynesian cosmogonic myth: 'in the beginning, there were only the primordial waters, plunged in cosmic darkness'. The supreme god Io 'expressed the desire to emerge from his repose. Immediately, light appeared.' Form is instituted in the act of creation: 'Then he went on: "Ye waters of Tai-kama, be ye separate. Heavens, be formed!"'[4] The material world itself is often described as emerging from a primordial state lacking definition and the properties of matter. From the Achomavi people of North America: 'In the beginning there was water everywhere and the sky was clear and cloudless, but suddenly a cloud formed in the sky, condensed and changed into Coyote.'[5]

The pre-existent state before form and matter was also a condition without time. For the Maya of Yukatan, 'in the year and the day of darkness... when the world was still sunk in darkness and chaos and the earth was covered with water', time did not yet exist: 'there were no days and no years'. Time, space, the world of matter and humanity come into being through an act of divine sexual union: 'one day appeared the Stag-God, who had the name of Puma-snake, and the beautiful Stag-Goddess who was called Jaguar-snake. They had human shape and great magical power; they united sexually and from that the world originated.'[6]

In some ancient mythologies, the initial void or formless state is given form or structure by a human quality or imperative. The cosmogony of the Rigveda states: 'In the beginning there was only water, and in the water was a living germ. Out of this living germ everything was born through Tapas' (desire). This first germ was the 'earliest seed of spirit' which brings 'the force of heat' to a universe 'indistinct and fluid', 'covered up by darkness'.[7] A Japanese creation myth narrates a similar primordial state containing the seed of the familiar world: 'Of old times the Sky and the Earth were not yet set apart the one from the other... All was a mass, formless and egg-shaped, the extent whereof is not known, which held the life principle.'[8] A Chinese creation narrative from the Han period describes the emergence of the finite material world from an earlier disorder: 'Before heaven and earth had taken form all was vague and amorphous... The Great Beginning produced emptiness and emptiness produced the universe. The universe produced material-force which had limits.'[9] These limits include the borders of the material world and the time and space experienced by mortal humans.

Mythological time and space

The conceptualisation of time and space in mythology entails a dual focus: there is the time and space experienced by the social group, but also, more profoundly, there is the time and space of the spirit-world. Hunter-gatherer societies narrated their belief in spirit-ancestors or spirit-creators in their oral traditions, handed down through generations of story-telling; the spirit-world was also given visual representation in art, which generally served a ceremonial purpose. This bifocal sacred/profane sensibility within traditional cultures relates directly to their understanding of time and space. The territory occupied or traversed by a particular tribe could be controlled by social rules and restrictions, but this geographical territory represented only one surface aspect of the space known to the group. Everywhere this territory was thought to be infused with spiritual presence. The spirits of ancestors were deemed to dwell in a metaphysical space adjoining the material space; the spirit-world thus animated the landscape. Totemic centres were perceived as sites of great positive spiritual energy, so that the land in many positions had a 'bipresence' in which two different registers of space mingled.[10]

Certain specialist individuals, such as shamans, were considered to have the power to traverse both spatial planes, for the benefit of the group. In order to cure an ill member of the tribe, the shaman self-induced a trance-like state, often through physical deprivation or the use of hallucinogenic drugs. In this transformed state of consciousness, the shaman was empowered to travel to the spirit-world, traversing great distances of spiritual space, engaging totem spirit-animals or ancestors, in order to find a cure for the individual dwelling on the terrestrial plane. The two versions of space – the worldly one encompassing the unwell individual and the spiritual one into which the shaman ventured – are two aspects of the same event, in a world characterised by dual presence.[11]

The spirit-world entered by the shaman constitutes another region of time as well as space. It was believed that the shaman's soul or spirit may journey through other worlds for a period of many years on its quest; this temporality, measured in earthly time, may correspond to a duration of only several hours. A Chukot Eskimo myth cited by Akhundov expresses a dual layering of time and space: a shaman travels through distant lands on different levels for two or three years, returning home to find that his children have grown old, as decades of local time have passed. Seeing his aged son, the traveller falls dead and disintegrates into dust.[12] This myth not only features the divergent temporalities of the two worlds, but also operates as a warning that for a human to leave his own time and space for too long ensures death.

Mythological time has other important features. The primary aspect is its orientation to the past, to the time of spirit ancestors. Narrated events of the mythic past exist in and explain the present, and can also guide the future. Rituals are performed so that an event from the mythical past may be acted out and kept alive; the present and future are 'perceived through the structure of the past'.[13] In the phrase made famous by Claude Lévi-Strauss, myths are 'machines for the suppression of time'.[14] The ancestral events continuously described through oral narration are understood not as 'history' – consigned irretrievably to the past – but as foundational events existing simultaneously in past, present and future.

Some appreciation of this multi-layered conception of time and space can be gained from a study of the art of Aboriginal Australia. Wally Caruana succinctly states the significance of art for Aboriginal people: 'Art is a means by which the present is connected with the past and human beings with the supernatural world.'[15] As practised for tens of thousands of years, Aboriginal art 'activates the powers of the ancestral beings'; it also 'expresses individual and group identity, and the relationships between people and the land'.[16] The meaning of Aboriginal art relates to the Dreaming or Dreamtime, the European terms for Aboriginal belief in a pre-historical state including the origin of the cosmos. In the Dreaming, creator ancestors such as the Rainbow Serpent travelled across the previously formless world, creating the land and everything in it while laying down laws of social behaviour. These events were depicted in artworks originally used for ceremonial purposes, often representing spiritual knowledge available only to initiated members of the social group; since the 1970s, Aboriginal artists have painted versions of Dreamings using acrylic paint and canvas, made available to general audiences and the Western art market.

A painting will refer to a specific Dreaming relating to the 'country' or territory known to the artist. Symbols within the painting may denote physical features of that territory, or they may represent traces of the ancestral beings of the Dreaming; one pictorial figure may symbolise both levels at once. The painting is thus a spiritual map of the territory, encompassing both its physical and metaphysical characteristics, both its present and its past. The inherent ambiguity of this form of representation means that interpretation of the painting will depend on the viewer's level of ritual knowledge and understanding of the Dreaming associated with the territory. The iconography of desert art conveys something of this dual register. A concentric circle may represent a site, camp, waterhole or fire; it may also symbolise the portal by which the ancestral spirit comes up through the earth and later returns to it. Straight or meandering lines may denote water courses or lightning; they may also trace the paths of spirit ancestors or supernatural beings. The spiritual map, unlike a conventional map, has more than one level of meaning, because it is depicting more than one register of time and space within the territory.

Cycles of time

Time in traditional cultures and in ancient religion was conceived not as a straight line, but as a cycle. The idea of progress, of time moving forward to an improved social state, was unknown in the ancient world, even in the great civilisations. This is not to say that innovation was impossible for the ancients; Roman engineers, for example, found ingenious solutions to many technical problems. But there was no sense that the present is better than the past, and that the future will be better than the present, which defines the notion of progress. Rather, time was thought to loop around in great circles in a process of endless repetition, and the period of greatest human happiness was generally deemed to reside in the past.

The cyclical nature of mythological time has been most thoroughly analysed by Mircea Eliade, in his study of 'the myth of the eternal return' and its many variant forms. For Eliade, the ritual act, in its repetition of gestures and archetypes, entails 'an implicit abolition of profane time', so that participants in the ritual 'are transplanted into the mythical epoch in which its revelation took place'.[17] This revelation often involves a creation narrative; Eliade finds many instances in which the recital of cosmogonic myths is part of ceremonies for cure, rebirth or creativity. The Navajos, for instance, recited their creation myth in connection to cures; through this ritual the Medicine Man is able to access the spirit-world and effect a cure. The patient is transported as well, as a result of the creation myth ritual, 'back to the origin of the world' in mythical time and space.[18]

Eliade finds a similar process in the Brahmanic sacrifice ritual, which is conceived as a re-staging of the creation of the world. The ritual act 'reactualizes the archetypal cosmogonic act', so that the present moment is made to coincide with the 'mythical instant' in 'the continual regeneration of the world'.[19] The ceremonial act is a willed regeneration not only of the material world – animals and plants, rainfall and fertility – but of time itself. The cyclical conception of time allows for this inexhaustible renewal. On each occasion of ritual regeneration, 'everything begins over again at its commencement every instant'. Eliade finds a strength and optimism implicit in the cyclical world-view, as 'no event is irreversible and no transformation is final'.[20] Time recurs like the seasons; it does not fly in one direction like an arrow; the eternal return 'reveals an ontology uncontaminated by time and becoming'.[21]

Ancient cultures construed the cycle of time as a series of ages, in which the golden age was invariably far in the past. Indian mythology held that the first age was the golden age, 'when virtue prevailed and man lived on the fruits of the garden of earth'.[22] The second and third ages entailed a decline in virtue and lifespan, while the advent of the present Kali Age brings the last and worst of the four ages, an era of sin and disease. The end of the Kali Age will be brought by fire and flood, destroying the earth; from the cosmic waters the cosmos will be created anew, heralding a new golden age. This great cycle of birth, decay and renewal will repeat until the end of time. The eastern religions further enshrined the cycle of time within the doctrines of re-incarnation and karma: in Hinduism, the soul works out its karma through a cycle of birth, death and rebirth, aiming for *moksha* or release from its *samsara* (wandering) throughout this cycle. In Buddhism, the practice of *dharma* offers the way to enlightenment, when the rebirth cycle of bodily existences is broken, karma is shed and *nirvana* – a pure spiritual state – is attained.

The ancient civilisations of Middle America conceived of time as a series of world-ages, each doomed to a cataclysmic end. For the Aztecs, the four ages were Water, Jaguar, Rain and Wind, while the present was a fifth age incorporating the sum of the others. Dread of flood and eclipse, thought to have ended the first two ages, was motivation for constant surveillance in some of these cultures, designed to ward off the end of the current age. In Toltec ritual, the efforts to prevent eclipse from consuming the sun and heralding apocalypse included constant sacrifice of human hearts and blood, so that the sun or Royal Lord could be strengthened for the continuing illumination of the world.[23] Time itself, in the form of the present age, is kept running by means of the sacrifice ritual.

The mythology of ancient Greece included an account of the Five Races of Man, a narrative of decline culminating in the present. The Greek cosmogonic myth, as related by Hesiod, has chaos as the initial state of the cosmos; sky and earth create the first generation of gods, the Titans. The Titan Kronos (whose name furnishes us with the word 'chronos' or time) separates sky and earth, allowing for the golden race to flourish on earth, free of disease and the need to work. The second, silver, race was created by the Olympian gods under Zeus, who overthrew Kronos; the last three races, beginning with the bronze, were all made by Zeus. The fourth race was the age of the heroes who fought at Troy, while the present age is peopled by the iron race, destined to labour and prone to trouble and sorrow. Hesiod is looking toward the next turn of the cycle of time when he laments his position within the iron race: 'Would that I were not among the men of the fifth race, but had either died earlier or been born later.'[24] From within this cyclical conception of time, the present is inferior to the glories of the past, and the future is actually a return to the past.

Ancient Greek time and space

The philosophical inquiry conducted in ancient Greece laid the foundation for much of the conceptualisation of time and space in Western culture. The earliest Greek philosophers launched investigations into the nature of matter and space, yielding ideas of the divisibility of space and the earliest atomic theory of matter. By the time of Aristotle, the world was understood to be spherical in shape; in the Hellenistic age, Ptolemy developed a form of mapping based on longitude and latitude. The writings of Aristotle and Ptolemy formed the basis of medieval cosmology for many centuries, while the mapping of space according to Ptolemy's co-ordinates continues today. Plato's theory of the Forms, an other-worldly realm of perfect space and eternal time, not only furnished a metaphysical dimension of Western philosophy, but, through later waves of neo-Platonist thought, infused early Christian theology and, much later, the Renaissance.

Greek philosophy generated a welter of competing analyses of the nature of time and space, but analytical Greek thought also inherited from mythology – as articulated in Hesiod's *Theogony* – a conception of dual registers of time. One was the cyclical time experienced by mortals (the time of Kronos), while the other was a 'higher time' of the gods, whose vantage point in eternity was outside of worldly time.[25] The divine perspective on time was static rather than dynamic, since the gods could see the future, which was locked into place by Fate. The human world could receive glimpses of the future through oracles (transmitted while in a trance or ecstatic state) or the various types of soothsayers; but the future could never be changed, not even by gods, as it was determined by Fate or 'necessity'. In the *Iliad*, Achilles learns of his death in an impending battle, yet marches into that combat, knowing that his death will not only occur, but already *is*, since the future already exists. The pre-Socratic philosopher Thales condensed this world-view into two statements: 'Wisest of all is time, for it reveals everything' and 'Most powerful of all is necessity, for it prevails over everything'.[26]

Greek philosophy dealt with time and space as abstractions amenable to rational analysis. Space could be divided into infinitely small sections according to Anaxogoras, a notion criticised by Zeno in his famous paradox, in which Achilles is doomed never to catch the tortoise because the space between them is infinitely divisible. Democritus posited that the atom, physically indivisible, was the smallest division of matter. The Atomists asserted that the 'uncuttable' atoms moved through empty space, while the Eleatic school refuted the idea of the void as a non-space. For the Pythagoreans, the essence of space was mathematical, in that every element in matter was construed as a number, existing within a unity or harmony of number within the Cosmos. Protagoras brought a human dimension to philosophical inquiry – 'Man is the measure of all things' – that included the idea that each society should create its own *nomoi* – laws, but also customs – that best suited them;[27] this is an early expression of the *nomos* as socio-cultural space.

If the nature of space was vigorously debated by the competing schools of philosophy, time was not neglected in these debates. Heraclitus argued for a dynamic time associated with the idea of the world as 'an ever-living fire, kindled by measures, quenched by measures'.[28] While this fire metaphor suggests that Heraclitus' dynamism was circumscribed by the cyclical concept of time (kindled and quenched according to circular repetition[29]), his metaphor of the river has become the most cited statement of flux and becoming in Western thought: 'it is impossible to step twice into the same river'.[30] Movement and change were contained, however, even for Heraclitus, within the orbit of Logos, which was the 'essence of Fate' and the 'measure of time's appointed circle'.[31] In contrast to Heraclitus, Parmenides emphasised that 'all reality is one', the One Being transcending motion and change. The perspective of the One is eternity, meaning that temporal progression becomes meaningless, as does space and motion.[32] Zeno's paradoxes of motion were in part devised to protect Parmenides' world-view from opposing arguments emphasising dynamism and movement.

Parmenides' conception of the One, as an immutable plane removed from the changeable world available to the senses, was developed by Plato in his theory of the Forms. The Forms are unchanging perfection, the models for all things on earth, which are all inferior versions of their respective Forms. Humans – particularly philosophers – could glimpse the domain of the Forms through the reasoning intellect, while Plato held that there was a supreme Form, that of the Good, which sustained the other Forms. Time and space for Plato could be understood only in the context of the changeless world of the Forms. Space was 'the nurse of all becoming', while time was regarded as the moving image of eternity.[33] That is, space was the material element in which events unfolded, and time on earth could be experienced and measured. But the world of the Forms was eternal, possessing all time at once, so that past and future were meaningless concepts in a world which always 'is'. The observable movement of the celestial spheres was for Plato only an imitation, in its circular form, of the immobile eternity of the Forms.

Plato's thought was the culmination of an 'ontic' or static conception of being also found in other Greek philosophers, including Parmenides and the Pythagoreans. From the perspective of an infinite, perfect metaphysical zone, movement and change were regarded as 'inferior degrees of reality'.[34] Being is elevated over becoming in this thought. True being resides in the timeless Forms, while the mortal world of change proceeds in a series of cycles, 'in which the same reality is made, unmade and remade, in conformity with an immutable law'.[35]

One significant factor in the development of Greek philosophy is also highly pertinent to a history of space: the invention of writing. Literacy spread through ancient Greece from the origin of the Greek alphabet in the eighth century BC; after a slow transition of three centuries, the philosophical inquiry of the classical age was – after Socrates, who wrote nothing – conducted in writing. Walter J. Ong, who regards the invention of literacy as the most fundamental shift in human consciousness, argues that its radical impact derives from 'the reduction of dynamic sound to quiescent space, the separation of the word from the living present'.[36] A written text is a new form of space: a space for knowledge. It is a container of knowledge, an artificial memory, whereas previously the knowledge and history of a people had been conveyed in the narratives of story-tellers and poets. The Homeric epics were vast narrative poems, committed to memory by poets, performing a central didactic role in propagating ethical values: Homer was considered 'the educator of Greece'. The great epics operated according to the requirements of oral composition and transmission; this necessitated certain attributes, including formulaic thought-processes and descriptions, repetitive and conservative expression, agonistic narratives, external rather than internal crises, a subjective rather than objective orientation: 'for oral cultures, the cosmos is an ongoing event with man at its centre'.[37]

Eric Havelock has argued that literacy was the 'preface to Plato', that Plato's abstract philosophical thought – even the idea of the Forms themselves – was conditioned by the properties of writing. Plato's proscription of poets in *The Republic* is interpreted by Havelock as a conflict between oral and literate sensibilities. Despite his professed hostility to writing in *The Phaedrus*, Plato is in fact fully shaped by the properties of literacy. His banishment of the poets is his rejection of the oral tradition, his attempt to replace Homer with philosophy as the true educator of Greece. Literacy enables the objectification of knowledge, even the objectification of the self as an object of study; it enables thought to conceive of concepts, such as the Forms, which exist outside concrete reality. In preferring logic to narrative, the abstract to the concrete, the literate Plato usurps the oral Homer as the educator of Greece.

It was Plato's pupil, Aristotle, who laid the philosophical foundation of Western thought's 'mature spatial and temporal metaphysics'.[38] Space for Aristotle was understood as 'an objective category, a property of natural things'.[39] The logical apparatus established in Aristotle's philosophy was the basis of scientific inquiry into the properties of the natural world, including the measurement and analysis of space. Aristotle's conception of space had a relational aspect as well: space was regarded as a 'result of object relations in the material world'.[40] Aristotle proved that the earth was spherical by observation and analysis of stars, the lunar eclipse and the movement of ships beyond the horizon. He developed a cosmological model, with a geocentric base and a series of interlocking celestial spheres, which formed the basis of the Christian and medieval world-view for almost two millennia, before its displacement by Copernicus' model in the sixteenth century.

Technologies of space: The map

No map is a perfectly accurate depiction of territory; it is only a representation, in two dimensions and to scale, of three-dimensional space. It is a selection of features within that territory; not everything in that landscape will be represented. For a map to be a truly faithful representation of space, it would need to be at a scale of 1:1, including every feature of the land at correct scale; this feat has only been achieved in fiction, in Borges' celebrated story 'On Exactitude in

Science'. In this short fiction, an empire's ambition to map the entire world yields a map identical in size to the territory it depicts – but this feat of cartography is later abandoned as useless. The gigantic map is neglected, left frayed and tattered stretching over its lands.[41]

There are many types of maps, and many possible histories of mapping. In the broadest sense, to map is 'to take the measure of a world' and to communicate that measure, as Denis Cosgrove – one of many scholars of cartography – has stated.[42] A map is an interpretation of space that bears the imprint of the world in which it is made. In the words of historian Brian Harley, cartography 'belongs to the terrain of the social world in which it is produced'.[43] In short, 'cartographers manufacture power'.[44] A map represents territory according to the knowledge and world-view of its culture and its period. This means that a map may be a mathematical rendering of space – but it may also be a spiritual or religious rendering of space. Cave paintings and rock carvings included representations of the landscape, while the Aboriginal paintings of 'country' described earlier may be understood as spiritual maps of the territory known by the artist.

The oldest surviving portable map is a Babylonian clay tablet dated c. 2300 BC, featuring cuneiform inscriptions denoting property owned by an individual citizen. A later (c. 600 BC) Babylonian world map is an early exercise in mapping-as-power: the Persians and Egyptians are omitted from representation in this map of the world. The development of mapping as a mathematical representation of space occurred in the Hellenistic period following the reign of Alexander the Great, from the third century BC at the Great Library of Alexandria. Eratosthenes, who from 245 BC was head librarian, was able to draw on thousands of papyrus scrolls detailing geographical information, as well as previous small-scale Greek maps going back three centuries.

Eratosthenes coined the word 'geography' and maintained that the world could be accurately measured and mapped. He calculated the circumference of the earth with remarkable accuracy, while dividing the earth into climactic regions, meridians and parallels. Simon Garfield describes Eratosthenes' achievement as setting the 'template for a new goal – the formulation of a precise and consistent map of the world'.[45] Christian Jacob pinpoints the significant factor in Eratosthenes' mapping as abstraction: the world and its locations were charted on a map defined by 'mathematical correspondences' derived from Euclid's geometry. Abstraction is central to the process leading from 'the empirical vision to the mental schematization' in representing global territory.[46] Jacob makes the further observation that the world map emanating from the library at Alexandria was itself an exercise of power, a 'process of intellectual control'.[47] A map displaying the territory claimed by Alexander the Great in effect re-claimed that territory through the act of mapping: 'its most remote parts were absorbed within a single homogeneous space, according to a single geometrical order'.[48] The map-maker effected 'intellectual mastery' of space through the appropriation of the earth and imposition of order according to lines, place-names and mathematical positions.

The other major development in cartography ensuing from research at the Library of Alexandria was the atlas of Claudius Ptolemy, produced around 150 AD. Ptolemy's *Geographia* contained co-ordinates according to a mathematical grid system – latitude and longitude – for about 8,000 cities and other world locations. In the *Geographia*, Ptolemy described methods for projecting the spherical earth onto a two-dimensional plane, while the mathematical co-ordinate system made possible both the mapping of localised regions and of the world itself.

Mappaemundi – maps of the world – were a prominent feature of mapping in medieval Europe, but such endeavours had very little in common with the mapping envisaged by Ptolemy. The maps made in medieval Europe were defined by the dominant religious world-view: they were imaginative illustrated constructions of both space and time according to the religious understanding of the period. These representations of the world were constructed with religious symbolism as their base: the body of Christ was often used as microcosm and geographical image, while Jerusalem was made the world's symbolic and geographic centre. Paradise could be represented on a map, but separated from secular territory, by ocean, wall or even fire: 'categories of time and space seem to be altered' when it comes to paradise in the medieval consciousness.[49]

In Europe, a return to mapping based on mathematical co-ordinates was made possible by the rediscovery of Ptolemy's *Geographia* in the late fourteenth century, after its preservation and transmission in the Arab world. Ptolemy's methods for rendering territory were combined, in the fifteenth century, with the invention of single-point perspective for representing space, and the development of printing. Renaissance mapping, unlike its medieval predecessors, was concerned with quantification and measurement: terrestrial space was represented in geometrical rather than symbolic terms. This methodological shift within mapping entailed a conceptual shift regarding the extended space of the world. World maps were constructed, as Jerry Brotton has observed, of the 'empty, homogeneous graticule of latitude and longitude', which meant that newly discovered lands could be inserted into the 'predetermined spatial grid'.[50] Maps were printed and then re-printed as they were modified by new discoveries: the European encounter with a new continent in 1492 'made each new map assume a historical and almost accidental dimension', as Alessandro Scafi has remarked.[51] This was also the year in which Europe's first known terrestrial globe – inspired by Ptolemy's theoretical models of projection – was produced.[52] As the age of discovery developed into centuries of European imperialist expansion, maps and globes of the world represented empires as world territory was divided up, often through the use of colour to designate far-flung imperial territories.

A change to the flow of time

A major shift in the conceptualisation of time occurred within Judaeo-Christian thought: this was the shift from cyclical to linear time. Mircea Eliade observes the centrality of time's cycle to mythological thought, in which 'the victory over the forces of darkness and chaos... occurs regularly every year'; in the ancient world, the cycle of time stipulated that although the current age may be inferior to the past, 'history could be tolerated' because of the inevitable return of the past.[53] In contrast, the first sentence of the book of Genesis is: 'In the beginning, God created the heaven and the earth.' The Hebrew Bible posits a definite beginning point for existence and for time and space.

This cosmological statement represents a break with earlier cosmologies built on the cyclical notion of time. Felipe Fernández-Nesto observes that by comparison with 'the beauty of a cycle, which has no beginning and no ending, Genesis proposed that time began with a unique act of creation'.[54] The Hebrew prophets 'discovered a one-way time', in the words of Eliade, 'transcending the traditional vision of the cycle'.[55] They did this by placing a value on historical facts, so that monotheistic revelation 'takes place in time, in historical duration: Moses receives the Law at a certain place and at a certain date'.[56] Revelation is not reversible and therefore becomes precious. The prophets projected their visions into the future, meaning that 'victory over the forces of darkness and chaos no longer occurs regularly every year but is projected into a future and Messianic' time.[57]

Christianity inherited this linear conception of time, with its focus on the future, from Judaism. Christians were instructed to look to the future, when the Second Coming would herald the Apocalypse and the end of time. Redemption of mortal sins held the promise of eternal paradise once the linear course of earthly time reached its end point. As articulated in the writings of St Augustine in the early fifth century, the Christian conception of time maintained a duality between eternity and the linear course of time experienced by mortals. Akhundov observes that in this regard Christian thought maintained aspects of the Platonic tradition: 'an absolute opposition between the static celestial God (eternity) and the dynamic earthly world (time)'.[58] Yet in another fundamental regard, time in Christian thought has 'reversed its course'. In mythology and Greek philosophy, time was construed as flowing from the past into the future. The new conception of time 'is turned toward a coming salvation and escape from time, so that the Christian river flows out of the future'.[59]

This orientation to the future within a linear conception of time had important consequences for later intellectual and political developments. The idea of progress, which emerged in the European Enlightenment of the eighteenth century, entailed a secularisation of the Christian concept of time. J. B. Bury, in his intellectual history of progress, credits Christian theology with laying the foundation for the idea of progress, in which the future is held to be a desirable state. For Bury, Christian thought brought a wholly new perspective to the course of human events, endowing that course with a 'definite meaning' and representing the past 'as leading up to a definite and desirable goal in the future'.[60] The spirit of progress celebrated the present as an improvement on the past, while looking with confidence to the future, which was assumed to constitute an improvement on the present. Enlightenment philosophers debated whether progress would reach an end point of social perfection, or would achieve an ongoing state of continuous improvement. In either version of progress, the orientation was to the future, and the conception of time was emphatically linear.

In the nineteenth century, the Christian model of historical time was the foundation first for Hegelian idealism, in which the World-Spirit sweeps to the end point of history as represented by the modern European state, and later for Marxist historical materialism. Albert Camus wryly observed the Christian underpinning of the atheist dialectical materialism: for Marx and Engels, the class struggle culminates in revolution – or salvation – and 'the final disappearance of political economy', the end of pain and suffering in history. At this point, the dictatorship of the proletariat will arrive, the state will wither and paradise will be attained.[61] While this parallel between Marxist and Christian narratives has been denied by many Marxists and has embarrassed others, some Marxist historians, such as Hayden White, have embraced the correspondence. For White, 'redemption' is key to both narratives, while the prophetic power of Marxism is rooted in the 'moral coloration that Marx derived from his Hegelian, utopian, and religious forebears'.[62] The secularisation of religious thought in the nineteenth century yielded these grand narratives of world history, as well as millenarian movements and revolutions. In the twenty-first century, when faith in progress has been severely curtailed, the end point of history is more likely to be associated with environmental disaster as a result of global warming: the future is no longer desirable, but is no less pressing.

Technologies of time: The clock

For traditional human societies, whether hunter-gatherer or based in agriculture, the experience of time was defined by the cycle of day and night, and by the rhythm of the seasons. Various means of marking the passage of days were practised before the invention of writing: these included inscriptions on cave walls, stone tools and bones. The megalith structures of the neolithic period, including Stonehenge, are thought to have been built in alignment with astronomical events such as the summer solstice. The invention of both writing and the positional number system by the Sumerians in Mesopotamia, by about 2500 BC, enabled the later inscription of calendars – the year divided into 12 lunar months – on clay tablets. Knowledge of the calendar – used largely for religious purposes of reckoning holy days – spread throughout civilisations after contact with Babylonian culture;[63] but there have been myriad forms of the calendar, each struggling with the actual length of the year (365.24219 mean solar days).[64] The Gregorian calendar, inaugurated by papal bull in 1582, was designed to prevent the date of Easter slipping too far, as it had in the Julian calendar; this new format has become the international calendar standard, although there are many other systems for organising the passing days, months and years.

If the calendar's smallest unit of time is the day, the clock allows division of the day into much smaller units of hours, minutes and seconds. (The division of the minute into 60 seconds and the hour into 60 minutes derives from the Babylonian numbering system, which was based on multiples of 60.)[65] Precursors to the mechanical clock included sundials, hourglasses, water-clocks and, in China from the eleventh century, an astronomical clock powered by hydraulics or flowing water. The motivation for the accurate measurement – and proclamation – of time was religious. In medieval Europe, the church structured the day and night into set times for prayer and chanting for monks, as marked by the toll of bells. The church also urged its congregation to attend church services often, drawn by the loud and regular peal of bells that came to punctuate the working day. The bells were regulated by sundials, hourglasses or water-clocks that told local time (i.e. shorter hours in winter, longer in summer). The invention of the mechanical clock at the end of the thirteenth century ushered in, according to Chiara Frugoni, 'a new kind of time, made of hours theoretically equal in length'.[66]

The new mechanical time was not necessarily precise: wear on the mechanism meant the lag of at least an hour over the course of a day. Frugoni remarks that the minute hand was not introduced until 1577, suggesting a degree of imprecision built into 'the broad tolerance that was accorded to the flow of time' in a society where 'most things were approximate'.[67] This was not yet an age of precision, strict meeting times or hard deadlines. But the process of regulating time was underway, as large public clocks moved from the church bell-tower to the town hall from the middle of the fourteenth century. These first public clocks had no dials or hands, as they simply sounded the hours; but their prominent existence in public space meant that 'time ceased to belong exclusively to God, and secular time was born'.[68]

The most significant aspect of mechanical clocks was their increasingly precise measurement of time, which was made evident to increasing numbers of citizens as clocks became more common-place in towns and cities. In his monumental study of 'technics and civilisation', Lewis Mumford declares that the clock, not the steam-engine, was 'the key-machine of the modern industrial age'.[69] Mumford regards the mechanical clock as the ur-machine that sets the template for industrialism: it was 'the outstanding fact and the typical symbolism of the machine'.[70] The clock made time quantifiable, regulated and regimented. Time was no longer variable, approximate, local or irregular: it had become abstracted and everywhere the same, as measured by clocks. The clock generated a new form of time, as Mumford notes: it is 'a piece of power-machinery whose "product" is seconds and minutes'.[71] Industrial capitalism became fixated with this new 'product' as a form of commodity, as reflected in the maxim 'time is money'. Time must not be wasted, it must be well spent; it must be strictly observed by clocks and watches, and it must be put to good – that is, profitable – use.

The sectoring of time and space

The development of visual perspective in the Italian Renaissance from the fourteenth century created the means of representing three-dimensional space accurately in the two dimensions of a painting. Treatises on linear perspective, such as Alberti's *On Painting*, published in 1435, were guidebooks for artists, codifying the space of the picture as a mathematical grid. This new mode of representing space entailed geometric projection based on a mathematical mapping of space. The scientific method cultivated techniques of observation and measurement, akin to a mapping of observable reality; Francis Bacon wrote in 1605 that his scientific approach has 'made as it were a small globe of the intellectual world'.[72] William Blake's famous painting of Isaac Newton depicted this 'paradigm of scientific reason' charting space – and human achievement – with a compass.[73] As Dennis Cosgrove has noted, the scientific revolution – and the Enlightenment it inspired – was waged with instruments of vision: the telescope, the microscope and the sextant. Modernity was distinguished by a concern for 'the human eye's physical capacity to register and to visualize materiality at every scale'.[74]

This is the sectoring of vision that has often been identified as a characteristic of the Enlightenment. It is the 'cartographic reason' or 'cartographic gaze' associated with scientific method, the age of exploration and mapping of new territories, and the instrumentalist approach to observing, surveying and measuring reality. The privileging of vision was evident in the philosophy of Descartes, who wrote in 1637 that sight 'is the most comprehensive and the noblest' sense, so that 'inventions which serve to augment its power are among the most useful that there can be'.[75] Scientific vision entailed detached observation and measurement of objects in space, as well as surveillance, calculation and prediction of behaviour in space and time.

Newton's three laws of motion and theory of universal gravity were formulated in his *Philosophiae Naturalis Principia Mathematica*, published in 1687, described by Stephen Hawking as 'the scientific foundation of the modern worldview'.[76] The *Principia* included Newton's concepts of absolute time and space: 'Absolute, true and mathematical time, of itself, and from its own nature flows equably without regard to anything external.' Likewise, absolute space, 'in its own nature, without regard to anything external, remains always similar and immovable'.[77] Newton regarded absolute space and time as constant throughout the universe, but beyond the perceptual limits of human beings, who perceived relative time and space, measured according to the movement of bodies. For Akhundov, relative space and time 'are the metricized analogs of absolute space and time'; that is, they can be measured in hours or metres.[78] Newton established time and space as 'the basic concepts of classical mechanics' and the 'ordering structures'[79] of theoretical physics.

One counter-force within Enlightenment thought to the empirical vision of time and space came in the philosophy of Immanuel Kant. In his *Critique of Pure Reason* (1787), Kant proposed that time and space are *a priori* notions, constructs of the human mind. These categories of the mind allow us to comprehend sensory experience. Kant's view has been upheld, to some extent, by later discoveries in the neuroscience of the late twentieth century, with major qualifications concerning developmental awareness of space.[80] The Enlightenment, however, remained inspired by Newton's concepts of absolute space and time, so that the clock became a symbol of the universe itself: ticking away reliably, and predictably, everywhere the same, according to universal laws.

If the conception of time and space was defined by Newtonian physics, the experience of space and time throughout modernity was affected by new technologies of transport and communication. Each of these contributed to a perceived shrinking of time and space as people, goods and information moved through space at increasing velocities. In her historical study of the postal service, Esther Milne notes that the advent of mail coaches in Britain in the 1780s increased the speed of mail delivery, prompting one writer to comment in 1787 that the mail coach had 'annihilated' time and space.[81] This perceived annihilation was in response to an increase in average speed from four miles per hour, previously achieved by postboy on horse-back, to nine miles per hour, achieved by mail coaches in the 1830s.[82] A more significant increase in velocity was attained by steam train. The first train carrying passengers travelled in 1825; the first to carry mail traversed the Manchester to Liverpool line in 1830. By 1840, trains averaged 23 miles per hour, improving to 30 miles per hour in the 1850s.[83]

Mail coaches and steam trains transported people and mail alike; a fundamentally new form of communication began with the electric telegraph in 1837. As Scott McQuire notes, the telegraph 'inaugurated the ability for messages to outpace messengers', redefining the practice of communication and the notion of proximity.[84] Prior to the telegraph, being present in one place meant being absent elsewhere. The telegraph created the possibility of 'action-at-a-distance', as the message travelled faster and farther than was possible for the messenger. McQuire observes that the idea of space informed by a dichotomy of presence and absence began to break down, as 'physically separated sites of action were bridged and juxtaposed in new ways'.[85] James Carey argues that the telegraph 'inaugurated new forms of knowledge production' by allowing 'symbols to move independently of geography' and of transport.[86] The telegraph was used from 1870 to regulate railway timetables in Britain; it was also connected to the stock exchange. Carey regards these uses of the telegraph as a 'fundamental discovery': the deployment of information as a 'control mechanism'.[87] The later communications technologies of the telephone (1876) and wireless radio (1894) extended the possibilities of information as 'action-at-a-distance'; global communication networks were built as information parallels to railway and sea routes.

One consequence of these developments was the demise of local time and its replacement by world standard time. A global framework was established by an international convention on telegraphy in 1875; this was followed in 1884 by the Prime Meridian Conference, which fixed Greenwich as zero degrees longitude and divided the world into 24 time zones.[88] Instead of each location observing its local time, every location in a time zone would observe the same time. This made the compilation of timetables much easier for long-distance travel by train or ship; it also meant, as McQuire remarks, that time and space were 'calibrated on a global scale',[89] with a global administration of time based on a centre in the West. In July 1913, the Eiffel Tower transmitted the first time signal to be relayed around the world.

Later in the twentieth century, international air travel radically diminished the time taken to travel throughout the world, while in 1964 the satellite made possible television link-ups in real time to all continents. Writing in the same year, Marshall McLuhan marked a culmination of 'more than a century of electric technology' dating back to the electric telegraph, 'abolishing time and space as far as our planet is concerned'. For McLuhan, communications technologies, including the satellite, ensured that 'the globe is no more than a village'.[90]

The final step in the global administration of world time was achieved in 1972 with the consolidation of International Atomic Time. This is a time-scale based on the averaging of several atomic clocks around the world; since 1967, the caesium atomic clock has been used as the standard device to measure time. A second is defined as 9,192,631,770 cycles of radiation corresponding to the transition between two electron spin energy levels of the ground state of the Caesium-133 atom. Co-ordinated Universal Time has been broadcast by radio stations around the world since 1972; today the satellite GPS is also used to synchronise time-keeping systems.

If the global ordering of time has developed in response to transport and communication technologies, the experience of time and space has changed over the decades since the advent of the telegraph. Scott McQuire considers the Industrial Revolution to have also been a revolution in the conception and experience of time, in which 'Newtonian physics legitimated the extension of mathematical time into virtually every sphere of existence.'[91] This included an internalisation by workers of a uniform social time kept by clocks and watches; E. P. Thompson's study of 'time, discipline and industrial capitalism' included the new social status attributed to watches, which could be bought by workers through subscription to 'watch clubs'.[92] The discipline of time was largely forged in the workplace, where the regulation and division of the working day emphasised the virtues of punctuality, precision and efficiency. Marx's analyses of the capitalist system of production pinpointed the buying and selling of time as central, as McQuire has observed: 'labour time is all the worker has to sell, while profit depends on the appropriation of the surplus time of others'.[93] The management of the workplace was also the control of an ever-more efficient regime of space and time.

The sectoring of time and space reached its extreme realisation in Frederick Taylor's *Principles of Scientific Management*, published in 1911. Taylorism was founded on the rigorous reduction of workers' tasks to the briefest possible allocation of time, to be repeated throughout the day. Taylor used time-and-motion studies of labour as the base for his simplified and highly repetitive system, which was implemented in the new assembly-line factories such as Henry Ford's automobile factory. The individual worker was reduced to the status of a cog in a vast machine; Taylorism was the blueprint for the 'iron cage' of rationalisation lamented by sociologist Max Weber as the inevitable consequence of modernity. It was the triumph of machine time within a vision of the workplace as a smoothly functioning mechanism. The human cost of this vision was displayed in Charlie Chaplin's satirical film *Modern Times* (1936). Chaplin plays the Taylorised worker, unable to switch off his mechanical task even in leisure time. Constant repetition has so internalised his machine-like behaviour that he has become a machine; the film's most famous image has Chaplin disappear into the cogs of the gigantic factory mechanism.

Spacetime

Space and time were considered separate entities in Newtonian physics; the new physics inaugurated by Einstein's special theory of relativity of 1905 considers them to be combined into a single manifold. One of the consequences of Einstein's 1905 paper was the overthrow of the Newtonian conception of time, as Stephen Hawking has observed.[94] Whereas for Newton 'absolute, true and mathematical time' was the same everywhere, Einstein proposed that all observers, regardless of how fast they were moving, would always measure the same speed for light. Relativity theory also maintained that an object's mass increases with velocity, an assertion later proven in experiments. Newton's absolute space, 'always similar and immovable', was a casualty of Einstein's 1915 general theory of relativity, which posited that a body's gravitational mass acts not only on other bodies but on space itself. A sufficiently massive body would induce space to curve around it; light would also bend around the object. This theory was proven in 1919 in an experiment during a solar eclipse, when light from stars was shown to bend around the sun.

Einstein's relativity theory came to symbolise a radical break in thought and in the conceptualisation of time and space. Although Newton's laws of time, space and motion continued to apply to machines, engineering and the movement of bodies on earth, they could no longer apply to massive bodies such as planets and stars; nor could they apply to the sub-atomic sphere opened up by quantum physics, which by 1927 had established that light could be construed as both particles and waves. Avant-garde artists such as the Futurists celebrated in 1909 that 'Time and Space died yesterday',[95] embracing relativity as a severing of the intellectual certainties of the past. The historian and scientist Gaston Bachelard, in a series of books from 1929, considered relativity and quantum mechanics to constitute a 'new scientific spirit' and a 'philosophy of no' in their rupturing of accepted configurations of knowledge.[96]

Although the concept of spacetime was implicit in Einstein's 1905 theory, it was made explicit in a 1908 publication by Hermann Minkowski, one of Einstein's teachers. Minkowski declared that, henceforth, 'space by itself, and time by itself, are doomed to fade away into mere shadows, and only a kind of union of the two will preserve an independent reality'.[97] Roger Penrose has observed that a 'four-dimensional space-time' as articulated by Minkowski is the key idea emerging from special relativity.[98] Space and time are joined and connected to matter and gravity. Changes to matter affect gravity, which can distort spacetime. Paul Davies pinpoints the shift from classical Newtonian physics to the concept of spacetime as involving a new perspective on time and space. In the Newtonian model, space was considered an 'arena in which things happened', while time was 'the backdrop against which the activity of these things was gauged'. In spacetime, time and space are no longer the backdrop: they are 'part of the play'. They can 'change, move, deform, distort'. Space and time are now construed as 'very much part of the physical universe'. This means that they have their own 'evolution, and maybe even their own birth and death'.[99]

Speculation on the birth of time and space is contained in the prevailing model of cosmogony in contemporary physics: the Big Bang theory. This theory proposes that the universe emerged approximately 13.8 billion years ago from a single point containing all space. The Big Bang signals the filling of the universe at an enormously high speed and energy level, including a phase known as cosmic inflation occurring after a miniscule fraction of the first second. Paul Brockelman, in his study of cosmology, notes that the expansion of the universe in the inflationary period is not so much an explosion from an initial 'bang' into empty space as it is 'an expansion of space and time itself'.[100] Time and space emerge with the Big Bang as dimensions of the universe; they are not, as they were for Newton, 'dimensions within which creation occurred'.[101]

Evidence supporting the Big Bang theory include the observation by Edwin Powell Hubble in 1929 that all galaxies are drifting apart at high speed, and the discovery in 1964 of cosmic background radiation, as predicted by the theory. The question of what existed before the Big Bang is left unanswered; or rather, it is considered unanswerable and is referred to as a singularity. Roger Penrose posited in 1965 that a black hole, formed by a star collapsing under its own gravity, would form a gravitational or spacetime singularity from which light could not emerge. A 1970 paper by Penrose and Hawking arguing that the Big Bang involved an initial singularity was generally accepted, although Hawking later reversed his view. The idea of an initial state of singularity implies, as Brockelman observes, that 'there was no "before" it, for space and time emerged at that moment'.[102]

Hawking has argued that since we cannot know what existed before the Big Bang, we should omit such consideration from the cosmogonic model and simply assume that 'time had a beginning at the big bang'.[103] The prevailing scientific theory of the origin of the universe thus has characteristics in common with the biblical account in Genesis. In the Bible, the universe – and with it time and space – simply comes into being, and there is no 'before' state. The Catholic Church in 1951 proclaimed that the Big Bang theory was indeed in accordance with the Bible.[104] This parallel between scientific and religious thought has driven other physicists in search of a more palatable cosmogony; various alternatives have been proposed including the multiverse and universe cycle models. One version of the latter was posited by Penrose in 2010 in his book *Cycles of Time: An Extraordinary New View of the Universe*, based on evidence in the cosmic radiation for the possible existence of an earlier universe before the Big Bang. Penrose argues that the Big Bang singularity is only an apparent singularity, and that the present universe was born as part of a cycle incorporating the death of a previous universe. One intriguing consequence of this theory is that it removes contemporary physics from the arrow of Judaeo-Christian time and returns it to the cyclical nature of mythological time.

2
Theorising Time and Space

My primary concern in this book is with the changes to the experience and conceptualisation of time and space apparent in the early twenty-first century. These changes correspond to the widespread use in developed societies of the internet, mobile phones and networked communications. Globalisation and the international flow of information are other related factors. Before considering the new time and space in detail, in this chapter I survey the wealth of theoretical approaches to the conceptualisation of time and space developed in the twentieth century, and into the twenty-first. Many of these theoretical perspectives pertain to the impact of technologies of communication and transport; others have elaborated analyses of the changing experience of time and space within modernity and postmodernity. The theoretical approaches considered in this chapter – drawn from a range of disciplines including philosophy, sociology, anthropology, geography, urban theory, media studies and technology studies – will serve as the base for my investigation of the contemporary condition of time and space in the next chapter.

Bergson's philosophy of time

Western thought in the first two decades of the twentieth century was heavily influenced by the philosophy of Henri Bergson, in particular his shaping of a new conception of time which he called *durée* or duration. Bergson was dissatisfied with the mechanistic world-view of Descartes and Newton, especially its mathematical concept of time. Duration was the centre-piece of Bergson's philosophy, which aimed to re-instate the lived experience of time, accessed more by intuition than by quantification and clocks. Bergson's thought exerted considerable influence in Western culture – in artistic circles as much as in philosophy – before falling into eclipse in the 1930s. Bergsonian ideas of time were revived in the 1980s by Gilles Deleuze, whose books on cinema – *The Movement-Image* and *The Time-Image* – drew heavily on Bergson. The time-based art forms, in particular performance art featuring extended durations of performance, have continued to draw inspiration from Bergson's formulation of intuitive time.

Bergson's initial training was in mathematics; he later credited his mathematical studies for provoking his interest in duration. He told the philosopher William James that his intellectual breakthrough came with the realisation that 'positive science essentially consists in the elimination of duration'.[1] Positive science or positivism was the systematic thought codified by August Comte in 1839, a rationalist generalised programme founded on mathematics and physics as the 'pure sciences'.[2] Bergson realised, to his 'great amazement,' that 'scientific time has no *duration*':[3] that is, no account of time as it is experienced, rather than as it is measured.

In his first major work, *Time and Free Will*, published in 1888, Bergson criticised the tendency in Western thought to objectify and quantify even the subjective experience of consciousness. Time is measured and counted, but this amounts to the spatialisation of time: 'when we speak of *time*, we generally think of a homogeneous medium in which our conscious states are ranged alongside one another as in space'.[4] What Bergson calls 'pure duration' is upheld by contrast as 'that which consciousness perceives', an 'intensive magnitude' that cannot be reduced to a quantity: 'as soon as we try to measure it, we unwittingly replace it by space'.[5]

Bergson developed his concept of pure duration in his major works *Matter and Memory* (1896) and *Creative Evolution* (1907). The former book contains consideration of the function of memory and the relationship between present and past. For Bergson, the past survives in the present in the form of motor memory, or the memory of the body as it repeats a movement by habit: repetition is the key to this performance of the past in the present. Bergson proceeds to question the status accorded the present, which is merely the 'indivisible limit' dividing the past from the future: 'nothing *is* less than the present moment'. The constant flow of time in fact means that 'we perceive only the past', so that the present is actually 'the invisible progress of the past gnawing into the future'.[6] The 'imaginary homogeneous time' as measured by clocks is alien to pure duration, which has multiple rhythms matching 'the degree of tension or relaxation of different kinds of consciousness'.[7] And memory acts to keep pressing the past into consciousness: 'memory does not consist in a regression from the present to the past, but, on the contrary, in a progress from the past to the present'.[8]

In *Creative Evolution*, his most well-known and influential book, Bergson expanded the conception of duration into a form of life-principle. Duration is 'the continuous progress of the past which... swells as it advances', so that consciousness, 'which is built up each instant with its accumulated experience, changes without ceasing'.[9] There is an emphasis on flux, creativity and becoming in duration: 'we are creating ourselves continually'.[10] Bergson contrasts this principle of becoming to the forms of Western thought in which time is 'theoretically negligible':[11] the Platonic tradition privileging being and stasis, the universal mechanism of science and the 'mechanism of our ordinary knowledge', which takes 'snapshots' of reality and strings them together like cinema.[12] Bergson's denigration of the new technology of cinema here was on the grounds that it spatialised duration in a mechanical manner; Deleuze, however, later salvaged the cinema, and its poetics of time, for a Bergsonian interpretation in his two books on the cinema-image. Bergson concluded *Creative Evolution* with a call for intellect to be supplemented by intuition as another kind of knowledge, necessary because 'on the flux itself of duration science neither could nor would lay hold'. Intuition enables us to see beyond the mere moments of time to the flow of time: 'it is the very flux of the real that we should be trying to follow'.[13]

Bergson's cultural impact was significant, particularly in the wake of *Creative Evolution*. The Italian Futurists, for example, were inspired by his emphasis on flux, becoming and self-creation, as were the cubists and French Symbolists.[14] Literary modernism, with its experimental techniques and exploration of interior states of consciousness, made explicit connections with Bergson's thought, in the cases of Gertrude Stein, William Faulkner, Wallace Stevens and others.[15] Marcel Proust, whose *In Search of Lost Time*, published from 1913, had the Bergsonian theme of memory and the endurance of the past, was personally acquainted with the philosopher (Bergson married his second cousin) and attended his public lectures. While Proust denied any explicit Bergsonian influence, his massive six-volume novel – which begins and ends with the word 'time' – is a literary parallel to the philosophy of memory and duration.

Indeed much of the modernist literature of the first half of the twentieth century was obsessed with the inner consciousness of time that Bergson called duration: the streams of consciousness in Virginia Woolf, James Joyce and William Faulkner articulate a subjectivity that cannot be counted in clock-time. Murad Akhundov finds the works of T. S. Eliot and Joyce reflecting a revolt against mechanical and historical time: they are 'saturated with nostalgia for the myth of eternal repetition and...for the abolition of time'.[16] One of the most complicated renderings of consciousness and the perception of time was made in Faulkner's 1929 novel *The Sound and the Fury*. In the opening section of this work, narrated by a man with the mind of a child, the present and past mingle in a prolonged stream of consciousness, as sense-perceptions in the present provoke vivid memories of the past. The novel further effects a dislocation of time as it is narrated by different characters at various historical points. Faulkner expressed a wish in 1928 that this highly complex novel could be printed in different colour types, each one representing one of the many time-periods within the text; an edition of the novel published in 2012 realised that wish, as two Faulkner scholars identified 14 different time-levels and accorded each one a specific ink colour.[17] The flow and sweep of time, the past gnawing into the present – in Bergson's terms – is artfully represented in this modernist literary work.

Bergson's influence on Western thought went into a swift decline in the 1930s. Part of the reason for this eclipse was sustained criticism from intellectual opponents: Bertrand Russell, for example, made a severe attack on Bergson's philosophy in 1914, rebuking its 'heaving sea of intuition' and accusing Bergson of disregarding rational thought.[18] Bergson's prestige was also dented by an ill-conceived engagement with Einstein, published in 1922, after which critics accused the philosopher of misunderstanding relativity. From the 1930s onward, there was little trace of Bergson for several decades within continental philosophy – dominated by Hegelian, Marxist and existential thought – or in the analytic tradition of Anglo-American philosophy. The 'return to Bergson' was largely effected by the philosopher Gilles Deleuze in the 1980s: Deleuze's afterword to the 1988 English translation of his *Bergsonism* advocated 'a renewal or an extension' of Bergson's project, 'in relation to the transformations of life and society, in parallel with the transformations of science'.[19] Deleuze interpreted Bergson as a 'philosopher of difference',[20] whose theorising of multiplicity, duration and embodiment could find ready application to contemporary issues and forms, including cinema.

Since Deleuze, Bergson's philosophy has been used as the base for analysis of computer-based new media,[21] while Bergson's concept of duration has continued to fascinate contemporary artists, particularly video artists and performance artists engaged with extended endurance (performance art and duration are discussed in Chapter 9). The Albanian-born, Berlin-based artist Anri Sala, for example, has made works in many media, including films that explore 'notions of durational and rhythmic time'.[22] Sala's sculptural work *Clocked Perspective*, installed in the Karlsraue Park at Kassel, Germany, for the Documenta (13) exhibition in 2012, confronted the differences between clock or objective time and subjectively experienced time. Sala's interest in the relation between 'time' and 'tempo' – even and uneven time[23] – is realised in this work whose mechanism has been specially engineered so that the large clock tells time despite its elliptical shape (see Figure 2.1).

This clock is a sculptural 'correction' of a mechanical clock incorporated in an 1825 painting housed in the Orangerie museum on the grounds at Kassel. In that earlier work, the clock was shown face-on, while its building was side-on; the break in perspective entailed by the inclusion of the mechanical clock in the painting is corrected by Sala's 2012 clock, which is shaped as if in perspective. The elliptical nature of the gigantic clock means that there are distorted distances between the hour marks – a stretching of the homogenised spatial time critiqued by Bergson. Observers of the strange public clock constructed by Sala are made aware of the arbitrary nature of clock time, as well as the mechanism that maintains its quantification of time.

Three theorists of space: Schmitt, Lefebvre and Bachelard

These three twentieth-century theorists of space differ widely in their political orientation as well as their disciplines, yet they each published original theoretical perspectives on space which exerted considerable influence on others. Carl Schmitt is perhaps the most

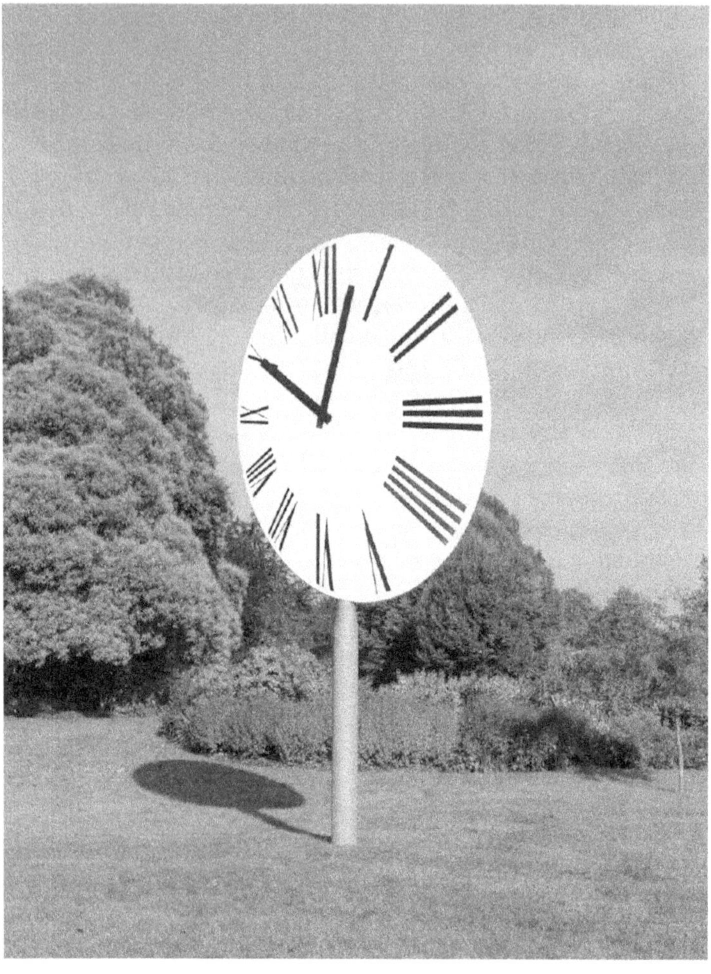

Figure 2.1 Anri Sala, *Clocked Perspective*, 2012. Painted aluminium and plastic clock case reinforced with fibreglass, metal clockwork and steel pillar. Clock: 400 × 280 × 30 cm. Dimensions when installed: 660 × 280 × 30 cm. Courtesy: Hauser & Wirth. Photo: Installation view at Frieze Art Fair, London, 2012 © Alex Delfanne

controversial inclusion here: a German political theorist and legal scholar, he was affiliated with the Nazis in the 1930s, ensuring that his work was stigmatised after the war. Yet his book *The* Nomos *of the Earth*, published in 1950, contained a world history and a meditation on the current global political order that was later found useful by a new generation of left-wing and radical political theorists. Central to this perspective was Schmitt's conception of *nomos*, by which he meant a spatial order or orientation to space specific to each society. Giorgio Agamben, for example, appropriated the concept of *nomos* for his own political purposes in his 1998 book *Homo Sacer: Sovereign Power and Bare Life*, in which he nominated the camp as 'the new biopolitical *nomos* of the planet'.[24]

In *The* Nomos *of the Earth*, Schmitt traced the meaning of *nomos* – or law – back to its ancient Greek origins as 'divide' and 'pasture': *nomos* was initially based, according to Schmitt, on land appropriation. The appropriation of land constituted 'the original spatial order', and was the source 'of all further concrete order and all further law'.[25] From this original linkage between law and land, Schmitt conducted a world history based on the premise that each society and each historical period is founded on its own *nomos*, understood as 'the fundamental process of apportioning space that is essential to every historical epoch'. This involves 'the convergence of order and orientation in the cohabitation of peoples' throughout world history. Every new age, nation, organisation of nations and empire, and 'power formations of every sort' are 'founded on new spatial divisions, new enclosures, and new spatial orders of the earth'.[26]

The *nomos* of the earth has constituted, at various times, a focus on land as territory, but also sea appropriations by naval powers or 'thalassocracies'. The Age of Discovery from 1492 altered the *nomos* of the earth as European powers competed to claim the 'free space' of the new world. A 'Eurocentric international law' was developed, based on 'the spatial order of firm land and the spatial order of free sea', which lasted for 400 years.[27] European public law gave way to international law after World War I; Schmitt nominates the 'spatial transformation of air war' as the other major shift in the early twentieth-century *nomos* of the earth. Air war brings with it the concept of territorial airspace, but also 'absolute disorientation' and a fatal blow to the old *nomos* based on land as territory.[28]

In a supplement to the book written in 1954, Schmitt observed the Cold War division of the earth into East and West, haunted by the fearsome prospect of nuclear destruction. He predicted that the winner of this conflict would oversee a new *nomos*, and that 'America is, so to speak, the greater island that could administer and guarantee the balance of the rest of the world.'[29] The post-1989 global *nomos* has indeed been constituted by America as sole superpower, challenged, since September 11, 2001, by the terrorist acts of radical Islamists whose vision of world order is the new caliphate. Other shifts to the *nomos* of the earth, in Schmitt's sense of spatial orientation, have been the colour images of the earth from space, first published in 1968, and images of environmental damage to the earth publicised widely by ecological movements. Mitchell Dean has argued that 'the great liberal utopia' of globalisation heralds a new *nomos*, 'an ordering of the earth and its beings which claims to do away with our orientation'.[30] The *nomos* has also been used as a theoretical term with a narrower sense, as in Agamben's designation of the camp – for refugees or political prisoners – as the new *nomos* of the earth: this is considered in Chapter 4, in the context of globalisation and displacement.

Henri Lefebvre's theoretical account of space comes from a different political perspective to that of Schmitt: Lefebvre was a member of the French Communist Party from 1928 to 1958 and remained a Marxist theorist after his expulsion from the party. In *The Production of Space*, published in 1974, Lefebvre's premise is that social space is a social product, and that each society, or each mode of production, produces 'a space, its own space'.[31] Lefebvre provides the example of medieval society and its feudal mode of production, which produced a space where manors and cathedrals were the 'strong points anchoring the network of lanes and main roads to a landscape transformed by peasant communities'.[32]

From this mode of medieval space grew the towns which were the spatial focal points of early capitalism. Advanced capitalism, however, produced 'abstract space', the world of commodities and the flow of finance, a space 'founded on the vast network of banks, business centres and major productive entities, as also on motorways, airports and information lattices'. Within this abstract space, the town – once the 'centre of historical space' – has 'disintegrated'.[33] Lefebvre finds no specifically socialist space to have been produced by the socialist states of the twentieth century, no 'architectural innovation', prompting him to speak of a 'failed transition' to socialism.[34]

Lefebvre's Marxist analysis offers a world history of social space based on the specific mode of production of any society. (E. P. Thompson proposed a similarly Marxist interpretation of temporality based on mode of production in his 1967 essay 'Time, Work-Discipline, and Industrial Capitalism'.) Lefebvre's study of space encompasses multiple aspects: physical, mental as well as social space; representations of space; natural space; and built environments. He is concerned with both the conceptualisation and the experience of space: space for Lefebvre is 'at once conceived, perceived, and directly lived'.[35] Much of his focus in *The Production of Space* is on urban space and the 'state management of space' within cities;[36] Lefebvre was an advocate for citizens' 'right to the city' and was involved in the May 1968 uprising in France. His work was an inspiration for later theorists of urban studies and human geography, while he continued to expand his analytical approach: a posthumous book *Rhythmanalysis* (1992) proposed a 'science of rhythms' in its dissection of 'space, time and everyday life'.

Gaston Bachelard's *The Poetics of Space*, published in 1958, differs from Lefebvre's analysis of social space, and from Schmitt's account of spatial order, due to its focus on intimate space and on the imagination. While Bachelard's early career was based in the history and philosophy of science, by the 1950s he had shifted his theoretical approach towards phenomenology, with its foregrounding of descriptions of experience. The goal in phenomenology was to reveal the meanings of objects and images for human observers, free of abstract or systematic rationalisations. Bachelard describes the shift in his approach in the introduction to *The Poetics of Space*: he has had to 'forget his learning and break with all his habits of philosophical research', in order to 'study the problems posed by the poetic imagination'.[37] His aim is 'to restore the subjectivity of images and to measure their fullness',[38] which involves a personal rather than objective interpretation of space. This results in a 'topophilia', or focus on 'the space we love'[39]– specifically the house and particularly our childhood house.

Bachelard's 'topo-analysis' is therefore of the subjective appreciation of space, rather than of its architectural or mathematical rendering: 'Space that has been seized upon by the imagination cannot remain indifferent space subject to the measures and estimates of the surveyor.' The image of the house is 'the topography of our intimate being',[40] so that the house is inhabited rather than neutral space. For Bachelard, 'our house is our corner of the world', and our childhood house is 'our first universe, a real cosmos in every sense of the word'.[41] The day-dreaming reverie celebrated by Bachelard includes memories of our earliest home, when 'we travel to the land of Motionless Childhood'.[42] Bachelard identifies the sense of security and warmth experienced in our first years within our first home as a 'being-well' or well-being enjoyed before we are 'cast into the world'.[43] Memory and reverie of this first inhabited space enables us to return to this initial state of being.

The intensely subjective interpretation of space conducted by Bachelard may have drawn primarily on poetry and imagination, but his book has had considerable and widespread influence. Bachelard's celebration of the poetical dwelling has inspired later generations of geographers and other theorists of space, as well as architects and designers concerned with the ways in which constructed spaces are inhabited and perceived. The distinction between space and place – or the subjective rendering of space – became prominent in a range of disciplines from the 1970s. The personal appreciation of places – including houses – as theorised by Bachelard in his 1958 work has become a theme not only of architecture and urban geography but of contemporary art (this is considered in more detail in Chapter 4).

Postmodern time and space

Fredric Jameson, one of the leading theorists of the postmodern in the late twentieth century, proposed that if time was the dominant of the modern, then space was the dominant of the postmodern.[44] Postmodernism was generally thought to succeed modernism in Western societies in the 1970s, corresponding to the shift from industrialism to information as the economic base of those societies. First observed in architecture, then generalised to other means of cultural expression, postmodernism was characterised by irony, appropriation of images and found materials, quotation, fragmentation, and a focus on the spectacle, simulacrum or surface of commodity and media culture. Art critics and cultural commentators generally agree that the postmodern expired as a dominant cultural logic around the mid-1990s, replaced by network culture, or – in art – simply 'the contemporary'. Yet aspects of postmodern culture – particularly the circulation of images and data and the centrality of media technology – remain pertinent in the twenty-first century.

The postmodern re-formulation of time and space was founded to a large degree on the potential of new generations of technology. The digital sampler and digital scanner were widely available by the mid-1980s; these devices enabled the ready quotation of sound and image, as well as the construction of songs or visual artworks as assemblages of fragments. The philosopher Jean-François Lyotard, who defined postmodernism as an 'incredulity toward metanarratives' and a preference for 'little narratives',[45] in 1985 curated a major exhibition at the Centre Pompidou in Paris called *Les Immatériaux*. This influential exhibition focused on the 'immaterial' nature of information or data handled by computers; for Lyotard, the ongoing transformation of knowledge into information was a key component of the postmodern.

Jameson considered postmodernism as 'the cultural logic of late capitalism'; in his analysis, space was more significant than time in the postmodern because the quotation of past styles and fragments evident in postmodern culture constituted little more than pastiche or 'blank parody'.[46] The postmodern obsession with space, however, was for Jameson evident in the centrality of the image, projected everywhere on the ubiquitous screens in cityscapes; in the experience of the 'space of the city itself – the renovated and gentrified posturban city';[47] in the new abstractions of finance capital and networks of information; and in the fascination with the spectacle. Postmodern space in this account is multi-layered, fragmented and saturated by media.

Several other theorists, urban geographers and cultural critics, made observations on urban space in this period, often with reference to consumer culture. Edward Relph noted the 'placelessness' of many urban environments in 1976, a theme developed by Marc Augé in his 1995 analysis of the 'non-places' of 'supermodernity'. The generic malls and shopping centres dominating cities and towns were construed as anonymous constructions of space, lacking local character or any defining difference. Celeste Olalquiaga surveyed the modern 'megalopolis' in 1992, finding a deliberate confusion generated by the layout of shopping centres and ubiquitous video screens, a disorientation duly remedied by the concrete act of purchase.

David Harvey found 'the condition of postmodernity' in 1989 to be characterised by a 'space-time compression', effected by technologies of transport, communication and information flow.[48] Writing in a pre-internet environment, Harvey argued that the transmission of images by satellite, as well as other global networks such as the telephone, occasioned the compression of units of time and space; time is seen to 'annihilate' space, as images and information from around the globe converge at any geographical point. For Harvey, the postmodern urban condition entailed both a quickening of time and a shrinking of space. Writing at around the same time, the sociologist Anthony Giddens proposed a 'time-space distanciation' characteristic of late modern societies in a globalised environment (Giddens opposed the term 'postmodernism'). For Giddens, the increasingly virtual nature of electronic communications enabled 'stretched' relationships between the local and geographically distant, as time separates from space.[49]

The most detailed – and idiosyncratic – treatment of technology, time, space and postmodern society was conducted by the French urban theorist Paul Virilio. Although he did not identify himself as a theorist of the postmodern, as did Jameson and Harvey, Virilio wrote many short, fragmented books – in postmodern style – throughout the 1980s and 1990s, with titles such as *Speed and Politics* and *The Aesthetics of Disappearance*. Virilio's principal concern is with the transforming powers of technology, including the transformation of the experience of time and space. The key element for Virilio is technological speed, whether the speed of weapons, information, media or transport. Virilio coined terms to encapsulate the modern tyranny of technological speed: we live in a 'dromosphere' or speed-space; our consciousness is 'taken by speed' as our media technologies create a 'picnolepsy', or set of frequent breaks. Our vision is that of a 'montage of temporalities', made of shorter and shorter fragments, conditioned by 'the technologies that organize time'.[50]

The consequence for Virilio of this 'dromocratic revolution' is that space is increasingly eclipsed by speed. 'History progresses at the speed of its weapons systems',[51] he avers in *Speed and Politics*; the Gulf War was less a war fought over territory than it was fought with the 'logistics of perception', with satellite imaging and the electronic transmission of images, data and computer graphics. There is a certain over-statement, even an apocalyptic temper, to many of Virilio's pronouncements, but he has developed a sustained analysis of the impact of technological speed on the experience of space. The 'strategic value of the non-place of speed has definitely supplanted that of space', he declares, adding that the 'possession of Time' is the new form of territorial appropriation.[52]

Virilio's critical writing extended into the age of the internet: in an interview conducted in 1995, soon after the advent of the World Wide Web, he responded to the 'infosphere', predicting that it would 'impose itself on the geosphere'. Networked connectivity and interactivity meant for Virilio a 'reduced world', in which 'real space' is shrunk to 'nearly nothing'. The speed of information did not herald for Virilio a liberation, but rather a new form of 'incarceration': 'interactivity is to real space what radioactivity is to the atmosphere'.[53]

Neuroscience time

Neuroscience is the collective term used to describe an interdisciplinary cluster of sciences focused on the study of the brain and nervous system. The field developed in the second half of the twentieth century by drawing on developments in cognitive science and psychology as well as the emerging computer science. Technologies including functional magnetic resonance imaging have been used since the early 1990s to monitor the brain's responses to stimuli, enabling study of the relation between perception, brain and consciousness. One of the findings of recent neuroscience research concerns the perception of time.

In his 2011 book *Incognito,* the neuroscientist David Eagleman argued that 'our sense of time – how much time passed and what happened when – is constructed by our brains'.[54] Studies of the electrical signals in the brain carrying information from the senses reveal that auditory and visual information are processed at different speeds; yet these differences are 'edited out' by the brain, allowing us to perceive sight and sound at the same time. This is because the brain has made the assumption that regarding the event of our perception – a person knocking on a piece of wood, for instance – sound, sight and touch should be simultaneous. Any differences in the processing of sensory stimuli are 'smoothed out' to provide the illusion of simultaneity.

Another feature of time and perception revealed by neuroscience testing is the delay – measured in milliseconds but a delay nevertheless – between perception and processing by the brain. Neural processing of an external event takes time, because 'signals move fairly slowly in the brain, millions of time more slowly than electrons carrying signals in copper wire'. This has the effect that 'your perceptual world always lags behind the real world'. If the event you perceived was a snap of the fingers, 'at the moment you perceive it, the snap has already come and gone'.[55] The conclusion to be drawn from these studies is that we live fractionally in the past; in this regard, neuroscience has vindicated the proposal of Bergson – in *Matter and Memory* (1896) – that 'we perceive only the past'. Eagleman has suggested that although 'living in the past may seem like a disadvantage', it is 'a cost that the brain is willing to pay'. This is because the brain is assembling a coherent set of perceptions, 'the best possible story about what's going on in the world, and that takes time'.[56]

Another facet of time with a Bergsonian hue – the subjective experience of time as contrasted to objective clock time – has also been studied empirically by Eagleman and other neuroscientists. Eagleman's fascination with 'time and the brain: how subjective time relates to neural time' (the title of a co-authored paper from 2005) stems from a childhood experience. As an eight-year old, he fell from the roof of a house, and felt time slowing down as he fell, including a moment of 'absolute calm' before he hit the ground. He has collected 200 accounts of near-death experiences, each with a description of time appearing to slow down at the moment of extreme danger. Eagleman's conclusion is that time is a 'rubbery thing' that 'stretches out when you really turn your brain resources on'.[57] Other experiments performed in the laboratory have revealed the 'malleability of duration', when respondents' perception of the time of a simple event – the flashing of a square on a computer screen – can be distorted by changing the size of the shape.[58] The differences between subjective consciousness of duration and objective mathematical time have been illuminated by this recent neuroscience research.

Network time and space

The internet as a web-based and commercial entity emerged around 1994. Its military origins as ARPANET in 1969, with only four host computers connected, developed by the mid-1970s into a network of networks – an internetwork or internet – for researchers, largely university-based. By 1986 it was necessary to introduce domain names – 'edu' for educational institutions; 'com' for commercial enterprises – but the major expansionary developments occurred from 1991. In that year, commercial traffic was permitted on the internet, while British computer scientist Tim Berners-Lee, and collaborators, delivered hypertext transfer protocol to link web pages and software called the World Wide Web to view documents. The web browser Mosaic was marketed in 1993, Netscape in 1994; Amazon was also founded in 1994 as an online store. Use grew rapidly: 1.3 million personal computers were connected to the internet in 1993; by 1997, there were 19.5 million. In 1998, more than 100 million email messages were sent daily; by 2007, over 1.2 billion users were connected to the internet.[59]

The first sustained critical study of the social impact of electronic networks was conducted by sociologist Manuel Castells in his three-volume study, the first volume of which – *The Rise of the Network Society* – appeared in 1996 (substantially revised in a second edition in 2000). Castells is not concerned simply with the internet: his wide-ranging study includes other forms of networks including trafficking, and the many aspects of globalisation including trade and crime, while a focus of his work is the global inequalities across international networks. But the internet – as a newly emergent global electronic network – has a key place in Castells' formulation of 'the network society'.

Castells coined a pair of resonant terms – 'the space of flows' and 'timeless time' – to encapsulate electronic networks' reconfiguring of temporal-spatial orientation. 'The space of flows' is contrasted to the 'space of places', or geographically positioned space. The network society for Castells is characterised by a new spatial form founded on flows – of information, capital, media. It is 'the material organization of simultaneous social interaction at a distance by networking communication'; a later 2007 study by Castells and others of mobile phone communication found that mobile communication 'greatly contributes' to the spread of the space of flows as a 'structure of our everyday life'.[60] Castells is careful to situate the virtualisation implicit in the space of flows within constructs of physical reality; he also emphasises that the space of places is not eradicated by flows: 'the space of flows is not a placeless space; it does have a territorial configuration related to the nodes of the communication networks'.[61] But it is the flow of information across networks that for Castells has become the dominant form of spatiality in network society.

Castells' emphasis on the socially defining significance of the space of flows has not been without critics. Shaun Moores, in his 2012 study 'Media, Place and Mobility', argues that Castells initially made too severe a distinction between the space of flows, available to elites and professionals, and the space of places, associated with subordinates unable to access flows of information. This bifurcation resulted in Castells' contention, in *The Network Society*'s first edition, that 'we may be heading toward life in parallel universes'.[62] Moores notes that Castells later revised this dichotomous view, writing in 2000 that 'the geography of the new history will not be made, after all, of the separation between spaces and flows, but out of the interface between places and flows'.[63] The revised viewpoint implies that the flows of information, finance, images, text and sounds permeate all levels of society, rather than merely that of the elite.

'Timeless time' is Castells' term for the increased flexibility of time as it is 'processed' in network communication and with the aid of digital networked devices. Timeless time is the 'desequencing of social action, either by the compression of time or by the random ordering of the moments of the sequence'.[64] Simultaneous communications may take place, dissolving linear sequence. Castells refers to 'the blurring of the lifecycle under the conditions of flexible working patterns';[65] timeless time also refers to a severing of the experience of time from traditional patterns rooted in the past. Whereas communication once required long passages of time while mail was delivered, emails or texts can now be delivered in seconds, and past, present and future 'can be programmed to interact with each other in the same message'.[66] Time is not bound by history but 'managed as a resource' for increased productivity; the result, in Castells' view, is the 'supersession of time' in network society.[67]

From 2000, theorists in media studies, technology studies, geography, sociology, anthropology and other disciplines focused on the impact of networked communication on the experience of time and space. Lee and Liebenau referred in 2000 to 'internet time' and its effect on 'people's perception of time, and the way time is organised'.[68] Email and other forms of online communication have encouraged the dissolution of the work–leisure time distinction, so that time management often entails working via the internet on evenings and weekends. Increasingly, mobile phone networks have come to be recognised as contributing to an altered perception of time and space. Writing in 2012, Tom Chatfield described 'wired time' as a new condition for 'most young people in the developed world', who are 'never without access to the protective bubbles created by devices like smartphones and tablets'. Chatfield observed that the daily default state for these individuals is 'to be "wired" into at least one personalized form of media'.[69]

In a 2009 study of the reconceptualising of time and space in the era of electronic media, Panayiota Tsatsou cautioned against the kind of theoretical over-statement that renders physical place irrelevant and the lived experience of time superseded. Network time is not ultimately 'timeless', Tsatsou notes, as lags in time, connection time and deadlines are still important. Similarly, she argues that virtual or mobile spaces do not 'abolish' the significance of space; rather, they 'alter the structural conditions under which people conceptualise and experience space'.[70] Tsatsou's preference is for an emphasis on the 'mediated' nature of time and space in the electronic era: time and space are both 'essentially compressed and significantly distanciated' – but not obliterated or abolished – by electronic media and networked communications.[71] She draws support from Roger Silverstone's 2006 conception of the 'mediapolis' as 'the mediated public space where contemporary political life increasingly finds its place, both at national and global levels'.[72] This electronically mediated arena allows communication to take place despite spatial and temporal boundaries that would once have proven restrictive.

Graham Meikle and Sherman Young's 2012 study of 'media convergence' – that is, networked digital media – considers the means by which convergent media 'make possible altered approaches to mediating our experiences of time and space'.[73] One aspect of this electronic environment is found in videogames' construction of space, which invites users to navigate and search, 'exploring the spatial environment in order to understand its rules'.[74] Another is the 'time-shifting' available to television viewers who record, shift and customise their viewing. Finally, convergent media are shown to foster an acceleration of culture, as in the operation of Google Instant, which returns search results before the user has finished typing the query. This online development, introduced by Google in 2010, is based on the premise that a user saves valuable time scanning a results page while typing. The time saved, according to Google, is significant: 'If everyone uses Google Instant globally, we estimate this will save more than 3.5 billion seconds a day. That's 11 hours saved every second.'[75] The industrial-era motto of 'time is money' has become, in the post-industrial networked environment, 'time is information' – which can, of course, be monetised.

3
'No-One Is Where They Are': Virtual Time and Space

Mobile networked communication fosters a form of virtual time and space, which is superimposed onto territorial space: this is the new time and space of the early twenty-first century. Social media conducted through smartphones and other devices has created a new form of virtual space, in which individual users are frequently engaged. Individuals connected to the network via smartphone, tablet or laptop may be physically present but are in effect absent, engaged in a virtual conversation. As a result, the physical world is increasingly overlaid by the virtual network sphere: space is an overlapping of the virtual onto the geographical.

Meanwhile, time is increasingly composed of interruptions, breaks and distractions, as multiple messaging and information sources compete for users' attention. Our time is more filled than ever with messages, information and constant interruption; even social media devotees frequently complain of having too little time. Time is rarely used for long sustained contemplation of a single task; in an era of multi-tasking and instant messaging, time has become a montage of fractures. The extreme remedy recommended for the time-stressed, information-overloaded individual is a digital fast: the complete removal of oneself from the infosphere. But this is usually, in the manner of a fast, only a short-lived and temporary exemption from the unrelenting world of networked information.

The dawn of the new time and space

When did this new regime of time and space take hold? The writer and cultural critic Rebecca Solnit nominates 1995 as the year 'human character changed again'; soon after this point, she contends, our lives became 'a constant swirl of information' and communication dwindled to 'peremptory practical phrases and fragments'. In this new order, data is constantly interrupted by other data, individuals feel an anxiety about keeping up or getting behind and a certain 'quality of time' has been lost, never to be reclaimed.[1] The year 1995 corresponds to the spread of the recently established World Wide Web, as the number of websites, online traffic and email communications escalated. But 1995 is too early a date for the reconstructed experience of time and space, as mobile phone networks – and smartphones – had not yet been developed.

The first smartphones – that is, mobile phones equipped with operating systems allowing internet access and other multimedia functions – were developed from 1997, and in 1999 Japan became the first nation with widespread use of smartphones operating over a wireless network. The popularity of smartphones in other societies increased from 2004, corresponding to Web 2.0 and the rise of social media, including Facebook. Internet access became a major feature of smartphones, a feature highlighted by developments in 2007–2008. Apple launched its iPhone in 2007, showcasing improved rendition of web pages on a mobile phone; in 2008 Google launched its Android operating system. From this point, smartphone use became near-ubiquitous in urban environments, while the phones themselves became sophisticated, even glamorous objects, housing high-resolution cameras, touch-screen technology and internet access from almost anywhere and at any time.

2008 is therefore a plausible date to mark the new time and space: smartphone production, driven by the popularity of the iPhone and the Android operating system, saturated the market. Tom Chatfield cites surveys of media use conducted over a decade to support this view. A 1999 survey of over 2,000 Americans aged 8–18 found that the average time spent using media for this age group was six hours and 20 minutes each day. Chatfield notes that those conducting this survey considered this amount of media use to be close to 'saturation'; it was thought that it was not possible to consume any more media in any one day. This conclusion was in fact supported by a survey in 2004 of the same age group, which reported only a two-minute increase in daily media consumption. However, another survey of the same age group conducted in 2009 yielded a massive increase – of over 20% – in daily media use to seven hours and 40 minutes. If the use of multiple devices was included, media consumption rose to ten hours and 45 minutes each day. Chatfield notes that the major contributor to this surge was the use of iPhones and other smartphones 'to consume old and new media alike': watching downloaded TV shows, listening to music, sending text messages, checking Facebook and other social media sites, and sending and receiving emails.

The conclusion, as Chatfield remarks, is that within that half-decade between 2004 and 2009, media consumption moved beyond the saturation of leisure time to the 'complete integration' of mobile media into the routines and activities of daily life. Smartphone use continued to grow, so that by 2015 there were an estimated 1.75 billion smartphone users worldwide, while in Western nations such as Australia eight in ten individuals owned a smartphone.[2] This portable technology promises a mobile entertainment-information-communication system that can be carried constantly by the user. This means, for young users, that a 'portable, personal supply of songs, videos, games, applications and social-media services is permanently on tap'.[3]

From around 2007, social theorists began to focus on the cultural significance of mobile phone networks, as the rapid diffusion of smartphone use became evident. Manuel Castells and collaborators focused on 'mobile communication and society' in 2007, observing that 'the advent of this technology, allowing multimodal communication from anywhere to anywhere where there is appropriate infrastructure, raises a wide range of fundamental questions'.[4] Rich Ling and Scott Campbell proposed in 2009 that 'spatio-temporal boundaries are negotiated in a mobile society'. Because mobile technologies permit multiple schemes of time – across social media, texting, email, music, video and other platforms – 'clockwise time' takes on a diminished significance for users absorbed in their phones.[5]

Also in 2009, Panayiota Tsatsou remarked that 'the almost instant transfer of messages' across geographical boundaries and through time zones had created new spaces – such as online forums – and new measures of time – as in Twitter.[6] A later volume of essays edited by Ling and Campbell, in 2011, reflected on whether the escalating use of mobile phones was 'bringing us together' or 'tearing us apart';

one contributor to this volume, Kathleen Cumiskey, focused on the degree to which mobile phone use in a public place can withdraw the user from 'the copresent situation' in physical space into the mediated or virtual interaction via the phone.[7] The political theorist Giorgio Agamben complained bitterly, in 2009, that the cellular phone had 'reshaped' the 'gestures and behaviors of individuals' in his native Italy, rendering the relationship between people 'all the more abstract'.[8]

The dissection of space

Networked computer-based technology enables the dissection of space as a field of information, in which geographical space is overlaid by virtual mapping and imaging. This is most evident in military technology, which draws on enormous research and development funding to develop technologies of surveillance, targeting and imaging. The F-35 Joint Strike Fighter jet, manufactured by Lockheed Martin and due to become operational in 2020, features extraordinarily sophisticated data-gathering and vision systems for the pilot. The pilot's helmet receives data from six electro-optical sensors mounted on the body of the plane; the helmet's visor is a display screen of observational data, millimetres from the pilot's face. Through this display screen pilots attain a 360-degree view to detect threats and determine options; this enhanced visual field is known as 'situational awareness'. In addition, virtual-reality technology bestows a form of X-ray vision, allowing pilots to see through the cockpit floor 'as if it were transparent.' The US Air Force has described this augmented surveillance capacity as 'a beautiful God's-eye view of what's going on', powered by the availability of 'a stunning amount of information'.[9] This new-generation technology, installed in the most expensive military project in American history, equips pilots with a technologically enhanced field of vision – as close to a 'God's eye view' as has yet been achieved. It is in effect a super-imposition of data-enriched visualisation onto the visual field.

Key components of the networked society – the digital computer, the internet and GPS – all have military origins, the latter two dating from the Cold War period. The satellites used for military purposes of surveillance during the Cold War had their military restrictions lifted by the US government in 2000, enabling the civilian use of GPS to escalate. In-car satellite navigation (or satnav) became a standard feature of new cars, as drivers demanded the assistance of satnav, either through visual display or spoken instruction: the Dutch company TomTom sold 1.7 satnav boxes in 2005, quickly expanding to 65 million in 2011. Drivers increasingly chose to be guided by GPS navigation; Google Maps and other electronic mapping systems performed a similar function on smartphones.

It has been observed that satnav mapping has affected the way travellers engage with the space through which they travel. When we make frequent reference to electronic maps on dashboards in the car or on phones as we walk, 'we tend not to look around or up so much'. It has become possible to travel long distances with little appreciation for the surrounds, since an abstracted version of the territory has guided us on the visual display. This has been noted, in terms of lament, as 'a victory for sat nav', but a loss for 'geography, history, navigation, maps, human communication and the sense of being connected to the world all around us'.[10]

The enormously popular Google Earth – a popularity evident on its launch in 2005, when demand over-taxed the Google system and reinforcement servers were needed – fused satellite and internet technology in the 'zoomable' rendering of geographical space. Users are able to zoom in from outer space, taking in the territory of the earth, then continents, nations, states, cities, suburbs or towns, down to the very house on the very street occupied by the user. Simon Garfield, in his history of mapping, wryly notes that the electronic mapping freely available on the internet and smartphones is a new form of 'Me-Mapping', in which the user is constantly placed at the centre of navigated space. The combined technologies of the internet and GPS have fostered a 'reshaping of our own spatial abilities'. Where once European maps placed Jerusalem at the centre, or Chinese maps were centred on Youzhou, now 'we each stand, individually, at the centre of our own map worlds'. We plot a route on a phone or in the car from ourselves – 'allow current location' – and distance is measured 'from where we stand'.[11] The reconstruction of space through electronic mapping thus has a user-generated – or perhaps solipsistic – element, placing the user as origin and centre of the journey through space.

Another Google product – Google Glass – attempted to further enhance the visual field through the over-laying of information.

This technology, launched to the public in 2014, is an optical head-mounted display, a tiny screen built into a pair of glasses. The wearer communicates with the internet through voice instruction, while text and visual messages are conveyed to the wearer on the screen. Rebecca Solnit has observed that Google promotional videos for the device hold out the promise of 'glasses that interrupt you constantly'. The glasses render vision of the external world onto 'a screen on which reminder messages spring up'. Solnit has questioned both the desirability of this super-imposition of information onto the visual field, along with its assumption that 'our lives require lots of management, and that being managerial is our highest goal'.[12] Applications designed for Google Glass include one for facial recognition, enabling the wearer to scan individual faces with the aid of the software: this has provoked objections on the grounds of privacy, along with other criticisms of the civilian uses of technology like Google Glass. The ability to dissect and analyse space and the objects – including faces – within it, is offered as an enticing feature of this technology.

The ability to analyse social space is available to mobile smartphone users without the need for military surveillance technology or Google Glass. Many geo-location mobile phone applications – or apps – offer the means of dissecting public space without the need for an enhanced visual field. Location-based social networking apps, increasingly popular with users, draw on the GPS of a mobile phone to allow its user to locate and identify strangers within a vicinity. Social networking apps such as Blendr, Grindr, Roamz and Foursquare operate as a form of 'X-ray vision', transforming the perception of social space. Where once the individuals within a particular space would remain strangers, these apps can identify 'all the people who are up for meeting someone new, all the people who went to a school near me, all the people who speak French'. Some apps focus on specific interests: Grindr allows gay men to find other gay men. The effect of these apps in general is to give 'us eyes where we didn't have eyes before',[13] enabling a new scrutiny of social space. Apps with specific engagements with public space have proliferated: there are apps identifying parking spaces; others notify users of the location of parking officers; one app provides details of beach conditions; another provides guided 'culture walks' through city streets. There are apps to locate buses, trains and taxis in both space and time: the user is informed where the vehicle is currently located, and when it will arrive. Public space is criss-crossed by these user-targeted renditions of space.

Public space is networked space

When the social theorist Jurgen Habermas defined the public sphere – 'private people gathered together as a public and articulating the needs of society with the state'[14] – he used the example of the eighteenth-century European coffee house. This newly opened venue was a site of rational and informed discussion, befitting the goals of the Enlightenment in the eighteenth century. Patrons of these cafés debated social issues of the day, informed by newsletters, pamphlets and other publications available at the site. For Habermas, these coffee houses were models of the ideal speech situation, of informed and critical discussion, held freely and in public. As has been noted by many commentators, there were severe restrictions placed on entry to this public sphere in the early eighteenth century: the coffee-house customers were white, male and generally members of the emerging middle class – hence Habermas' characterisation of the new public sphere as a facet of 'bourgeois society'. Nevertheless, the coffee house represented for Habermas the possibilities of the public sphere, where individuals can come together, interact, discuss, argue and collaborate, where they can be inspired by meeting with others.

Eighteenth-century observers of an 'internet café' of the early twenty-first century would be struck by the poverty of social discourse held within this modern version of the coffee house. Instead of lively discussion between café patrons, there is silence, punctuated only by the clacking of computer keyboards, as individuals engage with machines and stare at screens. If there is speech, it is of an individual speaking to a computer screen, engaged in a Skype dialogue with an absent other. The eighteenth-century visitors would find the lack of discourse dispiriting, unless it could be explained to them that patrons of this café are fully engaged in discourse or debate – with conversation partners physically located elsewhere.

This is telematic communication: co-present at a distance, often a great distance. One individual seated in front of a computer may be in virtual discussion with an internet forum or chat-room comprising members from around the world. The Skyping individual may be in communication with another individual – or group – on a distant continent. Individuals may take their place in a social media user-group comprising hundreds or thousands of others. Players of multiplayer online games may consider themselves members of an online community numbering in the millions.

Both of these institutions serve coffee to their customers – but they have little in common beyond that. The eighteenth-century café was abuzz with animated discussion between coffee-drinkers; the twenty-first-century version features one end only of virtual conversations, resulting in near-silence within the physical space of the café. The difference corresponds to the contrast between the public sphere and networked public space. The differences between these two spaces expresses the shift from a spatial model of civil society to a virtual model, encompassing co-presence at a distance as conversations are held across vast expanses of space. Since much discourse is now conducted within networked communication, the public sphere has expanded to include a virtual component. The public sphere resides not just in coffee houses but also in the cloud, across elaborate and expansive networks.

Mobile communication technologies further confound received notions of the public sphere. When an individual conducts a conversation via a mobile phone – in a public place such as a café, square, park, bus or train carriage – that individual is engaged in a private discussion within the public sphere. The other conversant is physically absent from the space: only the mobile phone user is able to hear the person on the other end of the line. But fellow citizens within the public space are fully aware of the private conversation – or at least one half of it. In this instance, the private is super-imposed onto the public (often to the annoyance of the fellow-citizens); the virtual space opened up by the networked communication overlaps onto the public space.

There are many such forms of this rubbing-together of virtual and physical spaces, when the private/virtual sphere is seemingly preferred by users to the physical engagement with other citizens in the public sphere. Another expression of this preference occurs when users of iPods (or other listening devices) disengage from public space by wearing headphones or ear-buds, preferring the internal soundscape offered by the device to the sounds of the outside world. At the same time, those users may be thoroughly engaged with their smartphones. That engagement may include virtual encounters with social media user groups, conducted as the users walk or travel through public space.

Social theorists and cultural commentators have discussed this public behaviour since it was first evident: one of the earliest analyses was made by Ichiyo Habuchi in his study of mobile phone use by Japanese youth. Writing in 2005, Habuchi observed the practice of young mobile phone users creating virtual networks of friends and intimates, from which outsiders were excluded. He described this practice as 'telecocooning': 'the production of social identities through small, insular social groups', which strengthened collective bonds between the users, but did not allow 'the entry of strangers into such collective cocoons'.[15] This telecocooning – like the audio-cocooning of the iPod user – may remove phone users from public engagement. As artist and theorist Maria Miranda has noted, such individuals may seem 'cut off from their immediate surroundings'.[16] At the same time, these individuals may be engaged in elaborate interactions with other users of the network – sharing images, texts, music, information or links with scores of other individuals on social media.

Some commentators have expressed concern regarding the insularity of the mobile conclave, and the resultant withdrawal from civil engagement. Kenneth Gergen, writing in 2008, lamented the 'undermining of face-to-face communities' due to the use of mobile phone technology. 'Cell phone technology favours withdrawal from participation in face-to-face communal participation', he wrote, warning of a shift from civil society to 'monadic clusters of close relationships'.[17] The cultural historian Robert Pogue Harrison wrote in his study of the cultural significance of the garden, also in 2008, that most of the students at Stanford University did not actually see the splendid gardens on campus. This was because the students were 'much more at home in their computers' than in the 'three-dimensional world'. Harrison lamented that many students, engaged with their smartphones and audio-cocooned as they walk around campus, 'invariably walk through a dramatic Papua New Guinea sculpture garden with their heads down, as if afraid even to glance at the sublime towering forms' around them.[18] He notes that the students do not lack curiosity, and appreciate the landscape features on view when they are pointed out to them, but in such instances they are seeing the gardens for the first time, even though they have walked through them many times. Harrison regretfully concludes that the features of the real world, no matter how beautiful, 'have disappeared behind a cloaking device, it would seem'.[19]

This tone of lament – rising to a pitch of furious indignation in Agamben's dismay at the sight of his fellow Italians hunched over their smartphones – has been countered by more positive appraisals of mobile communication by other scholars. The emphasis in these accounts has been not on what has been lost in civil society, but what has been gained by the use of mobile networks. Rich Ling, writing in 2008, argued that the use of mobiles has not only strengthened 'in-group dynamics' but re-shaped social cohesion in general: mobiles have 'renewed and recalibrated social cohesion, offering the ability to work out new relationships and to organize broad political interactions'.[20] This view was supported in 2011 by Gerard Goggin and Kate Crawford, in their analysis of youth culture's engagement with mobile communication. Mobile phones, supplemented by new apps, have created 'different forms of publics and counter-publics, both face-to-face and at a distance'. The stress in this account is on the 'different modes of connection' to community, rather than on the dwindling of one mode, such as face-to-face interaction.[21]

The networked mediation of public space has made the conceptions of place, community and communication more complex, as Varnelis and Friedberg observed in 2008 in their study of 'networked publics'. The 'always-on, always-accessible network', they remark, has altered our concept of place, 'linking specific locales to a global continuum and thereby transforming our sense of proximity and distance'.[22] Varnelis and Friedberg offer examples indicating a transformation in the ways space and community are conceived. Bricks-and-mortar retail stores have been seriously challenged – often to the point of closure – by online 'stores' such as Amazon. At the same time, social media sites have come to seem like actual places, especially for young users who spend many hours at virtual 'teenage hangouts' offering a sense of intimate community.[23] The World Wide Web, composed of virtual spaces, has what could be called a 'spatial imaginary', as Louis Menand observed in 2014. The perception of places on the Web as actual places to visit or inhabit is furthered by the use of vocabulary borrowed from real estate: 'site', 'address' and 'domain'.[24]

The overlaying of the virtual network onto the physical space of cities has created a new 'hybrid' form of space, as Scott McQuire remarked in 2006. The twenty-first-century public domain is no longer simply defined by 'material structures such as streets and plazas'. But McQuire notes that neither is the new 'media city' solely a 'virtual space of electronic media'. The public domain is best understood as 'the complex interaction of material and immaterial spaces'.[25] The friction generated within this new hybrid form of public space occurs when the virtual, electronic space is seen to impose itself too disrespectfully onto the physical space. Theorists including Gergen, Agamben and Pogue Harrison have lamented the withdrawal from civil society occasioned by the preference of users for the virtual; others including Ling, Varnelis and Friedberg, and McQuire point to the potential for new modes of communication and community arising from the new hybrid forms of mediated space.

Moving through hertzian space

The mobile phone is in effect a tracking device, locating its user anywhere in the world through GPS. In a previous era, before the widespread use of mobile phones, it was acceptable for an individual to travel overseas for at least six months, communicating with family or friends at home no more frequently than a phone call every two months, or perhaps a postcard sent every month. This form of 'disappearance' from the social grid is almost unthinkable in an age when mobile phone users are always 'on-call' and always available. It is expected that an individual will respond to phone, text or email messages instantly, no matter his or her location in the world. Indeed, failure to respond to messages occasions either alarm or suspicion concerning the silent individual: either something adverse has happened or – equally disturbingly – the individual does not want to be contacted. The pressure to respond and communicate via mobile smartphones has changed the way space is experienced in travel. The bearer of the smartphone is always located, always connected somewhere in the global network.

The knowledge that territorial space is criss-crossed by electromagnetic waves has affected the way we conceptualise space itself. Anthony Dunne described mediated public space as 'hertzian space' in 1999, referring to the frequencies of electromagnetic waves emitted by mobile phones and other electronic devices.[26] As Maria Miranda has observed, awareness of space as 'hertzian' can alter the way we occupy actual space in a city: we look for 'hotspots', avoiding 'deadzones' and seeking places with a strong signal.[27] The desire for connectivity via the smartphone re-defines urban space as a wi-fi grid, or a map of spaces offering strong signals and an opportunity to join the network.

Artists were quick to explore the potential of location-based mobile media technologies. The term 'locative media' arose from a media art workshop in Latvia in 2002, describing artworks that are 'spatially localized, and centred on the individual user'. This new 'geospatial' art, using mobile network technologies, aspired to 'a collaborative cartography of space and mind, places and the connections between them'.[28] Artists were regarded as pioneers in the creative use of these new technologies, 'important contributors to the reconfiguring of mediated and networked public space'.[29] The electronics and communication industries recognised the role of artists, sponsoring projects and prizes in electronic arts; as Miranda notes, this occasioned some friction within the international arts community. The concern was that this locative media art may amount to nothing more than content for corporate technology, 'R & D for a new generation of entertainment spectacle'.[30] As a result, locative media artworks have tended to avoid the entertainment or game-playing possibilities of the technology, creating instead more oblique or enigmatic interplay between virtual space and the surrounding material space. At times reference is made to the history of specific sites, so that the user of handheld mobile devices may experience aspects of the site's formation that may have receded from view.

Mobile media art, which may combine moving image, sound and text in a handheld device carried by the user through a designated space, often promotes a deliberate overlapping or blurring of spaces. One example of this was the work *34n 118w* by Jeff Knowlton, Naomi Spellman and Jeremy Hight: this was a 'spatial fiction' based in downtown Los Angeles for users equipped with a tablet PC and GPS device. Viewers became participants in the 'fictive space' through which they moved.[31] Another recent locative work is *After Banhof Video Walk* by the Canadian artists Janet Cardiff and George Bures Miller; this work was set in the old train station in Kassel, Germany as part of Documenta 2012. Cardiff and Miller have created several works that 'transform our perception of space through the innovative use of sound and image technologies'.[32] Their locative media works have avoided the trap of becoming 'content for corporate R & D' through their poetic, often oblique representations of space. These works are recognised as carving out a 'unique space between high art and those mechanisms of spectacle developed by the most advanced sections of the culture industry'.[33]

After Banhof Video Walk was available to be downloaded onto smartphone or media player, or visitors to the Kassel train station could borrow players with headsets. Users were invited to walk with the artists – as a virtual presence – through the station. The resulting experience was an overlapping of the immediate environment by the audio-visual material provided on the player. Participants watched events unfold on the small screen, experiencing these events deeply because they were situated in exactly the location where the footage was shot. This work hovers in the intersection of worlds in hybrid mediated space. As visitors follow the moving images, 'a strange confusion of realities occurs'. It is as if the dual aspects of public space – the material and the immaterial – are blended and confounded in this artwork. An alternative space is concocted, 'where reality and fiction meld in a disturbing and uncanny way that has been referred to as "physical cinema" '.[34] This 'physical cinema' is the creative realisation of contemporary hybrid space.

'No-one is where they are'

One of the most astute analyses of the social impact of networked spaces is by the psychologist Sherry Turkle, whose book *Alone Together* (2011) is the final work in a trilogy of books assessing the relation between human users and computers, online interaction and, finally, social media. A feature of Turkle's research for this book is its base in the experience of social media users: Turkle conducted 450 detailed interviews, 300 of them with children and teenagers. Turkle notes from her respondents the necessity to be 'always on'; mobile phones are always switched on, meaning that their users are 'always at work, always on call'. Turkle's research concerns the ways in which 'we are changed as technology offers us substitutes for connecting with each other face-to-face'. From her interview subjects, Turkle learns of the 'many new kinds of space' made possible by the internet and mobile networking: couples who text each other while in bed together, so that their messages can be 'on the system'; family blogs where members construct an online space for their family; games and virtual communities where avatars represent the self in alternative spaces.[35]

The general observation made in *Alone Together* is that mobile phone users are perpetually tethered (to their phones) and marked absent (to physical space). The phone is a portable device that acts as a portal to people and places; but the user, while present to the individuals connected to the portal, is absent within his or her surroundings. As a result, a train station, café or park is no longer a communal space; people may come together 'but they do not speak to each other' – they are on their phones.[36] Turkle's experience at a conference where everyone was constantly on their smartphones or laptops, even during conference presentations, led her to the conclusion that 'what people mostly want from public space is to be alone with their personal networks'.[37] Family life is affected through the constant use of mobile phones and other devices by both parents and children. Parents complain that their children are not present at the dinner table when they are connected to their phones, tablets or videogames; yet Turkle also records many complaints from children aged eight to teenage that they are unable to secure their multitasking parents' attention. Parents push their toddler on the swing while speaking on the mobile phone; they email at the dinner table.[38] There is occasionally regret at the absence in physical space occasioned by the tethering to virtual space: as one respondent sighs: 'No-one is where they are. They're talking to someone miles away.'[39]

Efforts have at times been made to curtail the use of mobile technologies in public space – or at least in small sections of it. A girls' school in Perth, Western Australia, announced in 2013 that it had banned students from using mobile phones in the playground, due to concerns that the phones were favoured over face-to-face conversation. The school principal stated that 'being fully present in the moment, listening and speaking to one another and actually engaging in the art of conversation is a social skill that we wish to instill'.[40] Attempts such as this to restrict mobile phone use often appear as rear-guard efforts to redress social trends – but the move to ban mobiles at this school was understood elsewhere in the secondary school sector. The president of the New South Wales (NSW) Secondary Principals' Council noted the common sight of 'groups of students sitting together, each one of them immersed in a mobile device that is almost isolating them from the group they are sitting with'.[41]

Quiet train carriages, in which commuters are asked to refrain from talking on their mobile phones, were introduced in Queensland in 2010. The move to introduce a similar mobile-free space in NSW was welcomed, at least by writers of letters to newspaper editors: 'Relief from self-promoting galahs jawing endlessly into their accursed mobiles would be welcome.'[42] Since the readers of print newspapers tend to be older citizens, this forthright view on the violation of public space by mobile technology probably needs to be placed in an age-specific context.

However, a similar antipathy to the use of mobile phones in public circumstance has been repeatedly voiced in a youth context: rock concerts. Many musicians, promoters and music writers have complained of the practice by concert-goers of filming concerts using their camera-phones. The video, often of poor quality, is later uploaded to YouTube or elsewhere on the Web; but the actual experience of the concert has been diminished, as the user is busy recording the event rather than listening to it. Meanwhile, other concert-goers are annoyed by the camera's intrusion into their space and enjoyment of the music.

Some musicians have become so exasperated by this practice that they have attempted to ban mobile phones from their concerts; others have angrily seized and destroyed the devices.[43] In response to this 'peculiarly anti-social effect' of mobile phone overuse, a technology called Yondr was launched in San Francisco in 2014, offering the means of preventing phones from 'ruining concerts'. Concert-goers are given a slim case in which they place their phone; when in the concert area, the case locks, preventing users from accessing their phone. If users feel the phone vibrate, they can exit to regain access to their device and check their messages. The founder of this device, Graham Dugoni, created Yondr because he was convinced it was not possible 'to fully experience something and be documenting/broadcasting it at the same time'. The live show in a concert space is 'about being swept up into a shared mood'; when someone prefers to record the concert, they withdraw from the moment and shared experience. Psychologist Linda Henkel supported this view, stating that 'external memory devices' detract from the 'cognitive processing' of memory formation.[44] That is, the individual who recorded the concert is likely to remember little of experiencing it. In these mobile-free spaces, at least, something of the pre-mobile network sense of communal public space is preserved.

Network time

What of time in mobile networked culture? The common complaint is that nobody has enough time any more. Sherry Turkle's conclusion from interviews with social media users is that networked computers 'keep us busy'. Individuals may feel 'overwhelmed by the volume and velocity of their lives', and 'turn to technology to help us find time'. Those individuals are likely to find, however, that 'technology makes us busier than ever and ever more in search of retreat'.[45]

Mobile phone users are always on call, which means for many that they are always at work, ensuring a new definition of the work day, and of time. Studies of work–life balance have found that mobile networking has leapt 'the spatial boundary of the workplace', resulting in severe time-pressure for some workers, particularly full-time women workers with children. The 2012 work and life index report on Australian workers – entitled *The Big Squeeze* – found that 'at least a quarter of Australian workers are badly affected by work–life interference'.[46] The report noted that seven out of ten full-time women workers often or always felt 'rushed and pressed for time', while 41% of mothers with full-time jobs claimed they would prefer to work part-time.[47]

Mobile technologies have altered the locations of work, so that many workers will answer their emails even while preparing dinner at home. Perpetual connectivity brings with it an expectation that a worker will always respond to a message, even in leisure time. Messaging has escalated in volume, and each message sent in the 'instant response society' carries the expectation of a response, 'no matter what time it is, or even what day it is', as one worker ruefully commented.[48] The experience of many individuals of being short of time results in large part from the demands of constant communication and messaging made possible by mobile networking. 'I'm constantly pressed for time', laments another worker, 'I never have enough. I wish I could just stop the world and hop off and have some time to myself.'[49]

Teenagers reported to Turkle that they sleep with their phones, which are deemed so much a part of their identity that they have become 'like a phantom limb'.[50] Several surveys have found that users would rather give up sex than give up their mobile phones.[51] These devices, which many owners find indispensable, keep those owners occupied with a varied and constant stream of messages and information.

A typical user of a smartphone may, in any given hour, send and receive text messages, email and phone messages; upload or view images on Instagram; conduct text conversations on several social media sites; check Facebook status updates; receive a feed of Twitter messages; send and receive messages on Snapchat; conduct online shopping on Amazon or other commercial sites; bid for items on eBay; perform internet banking; watch a video on YouTube; access Google Maps; download music; check an entry in Wikipedia; search for information on Google; receive news feeds or customised information feeds from online news agencies or news filters; and access services and information using smartphone apps. Teenagers are unlikely to use phone calls, emails or Facebook – regarded as old persons' media – but are likely to log a high volume of text messages as well as messages within social media user groups. If the user is reading a text – a news item on an online news site or an ebook on a tablet – all of these other information sources may vie for attention.

The typical social media user feels pressed for time due to the sheer volume of messages. Because the smartphone is carried everywhere and at all times, there is no escape from the barrage of messages and information – and no escape is desired, in most cases. Because connectivity is so valued, new devices released to the IT consumer market have as one of their goals the enhancement of a user's connectivity. The Apple watch, launched in 2015, wirelessly connects to the wearer's smartphone, alerting the user by a vibration on the wrist when a new message has been received. Google Glass, which failed as a consumer item when first launched in 2014, had a similar motivation of keeping the wearer in touch – by visual means – with the stream of incoming messages. The overall aim of these assorted devices seems to be the filling of time with a constant flow of messages.

The streaming of audio-visual information on the internet and onto mobile devices adds another dimension to the constant flow of messages. On-demand viewing, whereby viewers can access programmes whenever they require – and for as long as they desire – has disrupted the stable time-ordering of scheduling, which was once the means of providing television entertainment. The regular programming of broadcast content, which began as a cultural form with radio broadcasting in the 1920s, and assumed a pattern of stability and repetition in television programming, is disappearing as a preferred presentation of programmes.

When digital TV programmes are archived by networks and made available via streaming at any time, viewers are increasingly unlikely to wait for an allocated scheduling of that programme. Digital TV viewers already have the means of pausing 'real-time' television, using the recordable hard-drive to fast-forward or rewind televised content: the viewer has an unprecedented command of television time. Content on-demand via streaming means that masses of content are archived and ready at any time – for 'binge viewing' or viewing in whatever order is preferred. The managing director of ABC (Australian Broadcasting Corporation), Mark Scott, said in 2015 that the reality of contemporary and future television viewing is glimpsed in a three-year-old seeking the cartoon *Peppa Pig* on an iPad. There is no sense of looking at a schedule and waiting for the designated time; there is instead the demand for the content now and the expectation that 'the technology will work seamlessly to deliver' it. Scott added that, in the wake of streaming and digital archiving of content, the schedule – with its weekly blocking of time and expectation that viewers would wait for the allocated time – 'will be one of those things that people look at in bewilderment'.[52]

Montage of fractures

The relentless stream of information – texting, email, phone messages, Facebook updates, Twitter and other stimuli – means time is fractured by constant interruption, perpetual distraction. In the 1980s, Paul Virilio wrote of the 'picnolepsy',[53] or frequent breaks structuring each consciousness exposed to contemporary media. This analysis related to pre-internet media; Virilio focused on the high-speed editing of images and sounds in television advertisements, or videoclips, as evidence of the rapid breaks and interruptions of which electronic media programming was composed. The age of the internet and social media, conducted throughout day and night on smartphones, constitutes an intensification of this condition of picnolepsy. There are now many more forms of message – most of them short – interrupting other messages in a perpetual play for users' attention. Time for these smartphone users has become a montage of fractures, a parade of breaks and interruptions.

The young subjects of Turkle's interviews acknowledge this condition; for many of them, interruption is the norm. An incoming message will always be accepted, interrupting an existing message because – as one interviewee insists – 'I have to see it. I have to.' Many smartphone users are gripped by this compulsion to check their messages. Some admit to texting while driving, such is the overwhelming need to connect. Others have suffered accidents in the street while engaged in messaging on their phones. When asked if they ever feel the need not to be interrupted, the young respondents do not understand the question, or the premise. 'I'm waiting to be interrupted right now', one responds. Turkle concludes that for these prolific social media users and texters, interruption is merely the beginning of connection.[54] A 2013 study found that smartphone users check their device on average 110 times per day.[55]

A battalion of neuroscientists, psychologists and media scholars has warned of the detrimental impact of this 'cognitive overload' of messaging on comprehension and other cognitive skills. Nicholas Carr assembled evidence from studies in cognitive science and psychology in his book *The Shallows* (2010); his conclusion was that the 'multi-message' format of web-based reading impedes the cognitive activities of reading and comprehension. A new form of reading has emerged, according to Carr and the specialists he cites: a superficial scanning of information, prone to distraction and interruption. He argues that the risk of this constant overload of information is that we become 'pancake people', in the words of playwright Richard Foreman: 'spread wide and thin as we connect with that vast network of information accessed by the mere touch of a button'.[56]

Certainly there are 'pancake people' who dismiss such claims as alarmist. Sam Anderson wrote an article entitled 'In Praise of Distraction' in 2009, in which he pointed to the centrality of connectivity in both work and leisure;[57] the ability to navigate through the many sources of information is a valuable skill. Many of the young social media users interviewed by Turkle – 'pikelet people', perhaps – revel in their ability to switch quickly between multiple modes of information. Texting – the preferred mode of teenagers – is often conducted as 'rapid response' messaging, while teenagers regularly juggle this mode of communication with shopping, social media, games, videos, music and other forms. Turkle notes that in the networking era, multi-tasking – which was once viewed as a 'blight' – has become recast as a virtue. Young people are 'lionized' for their ability to do many things at once, and multi-tasking is lauded as the crucial skill for the new digital culture.[58]

Here again, however, the evidence from science does not support this optimistic view. In *Distracted*, her 2008 book on the 'erosion of attention', Maggie Jackson undermined the claims for multi-tasking as a social virtue. Jackson cited many studies showing that switching between only two tasks added to the cognitive load and impeded thinking.[59] Turkle makes a similar argument, citing multiple studies in psychology, where 'multitaskers don't perform well on any of the tasks they are attempting'.[60] Earl Miller, a neuroscientist at MIT, goes further in stating that our brains are 'not wired to multitask well'. When individuals think they are multi-tasking, they are instead simply switching rapidly between tasks, with a 'cognitive cost in doing so'.[61]

The joy of messaging

Neuroscience adds a further insight into multi-tasking – and messaging in general – by focusing on the chemical reactions within the brain associated with sending and receiving messages. Neuroscientist Daniel Levitin reports that each activity such as sending a text message or checking email 'tweaks the novelty-seeking, reward-seeking centres of the brain, causing a burst of endogenous opioids'. This chemical response makes the individual feel good, but this is to the detriment of staying on a task. Instead of reaping rewards from sustained effort on a single task, in multi-tasking we reap 'empty rewards from completing a thousand little sugar-coated tasks'.[62] Turkle draws a similar conclusion from studies in psychology: 'multitasking feels good because the body rewards it with neurochemicals that induce a multi-tasking "high"'.[63] This reaction deceives multi-taskers into assuming that they are productive – but this is an illusion. Levitin adds that multi-tasking soon leaves the brain depleted of nutrients: switching from one task to another 'causes the brain to burn through fuel so quickly that we feel exhausted and disoriented after even a short time'.[64]

Texting has also been analysed by neuroscientists. As the primary mode of communication for teenagers, texting frequently activates the novelty centres in the brains of young phone users: each text sent 'delivers a shot of dopamine' as the limbic system's reward for having completed a task. The need to respond upon receiving a text plays into this system of expectation, task and reward, but as Levitin cautions, it is 'the novelty-seeking portion of the brain driving the limbic system that induces this feeling of pleasure'. If conducted on a repeated basis, this behaviour of constant checking of text or Twitter or social media constitutes 'a neural addiction'.[65] Turkle is wary of speaking of addiction as anything more than a metaphor, but acknowledges the craving to connect in young users, their nervous systems excited by the sending and receiving of messages so that they are 'stimulated by connectivity itself'.[66]

The sheer pleasure in sending and receiving messages is a significant factor in contemporary communications. Many of the phone and text messages sent – in their millions – are strictly unnecessary: 'Hi, I'm on the train and I'll be there in five minutes.' Before the advent of the mobile phone, there was no place, and no need, for messages such as these; there is still no need, as the sender of the message will meet the recipient in five minutes whether the message is sent or not. One facet of this message is its location of the sender in space – on the train – and time – five minutes away. But the primary function of this type of message is a phatic one.

Phatic communication is communication for its own sake, for the sake of keeping the channels of communication open. The term 'phatic' was coined in 1923 by the anthropologist Bronislaw Malinowski, to denote 'a type of speech in which ties of union are created by a mere exchange of words'.[67] The point of a phatic message is to express sociability and the willingness to communicate, to promote human warmth; in other words – to connect. The content of the message is less important than the existence of the message itself. The information conveyed may be negligible, but the phatic message, as a form of social talk, fulfils the social role of maintaining communication links. Many of the messages – phone and text – sent across mobile networks are of this phatic order; many of the messages consuming senders' and receivers' time are strictly unnecessary – but enjoyable – communications.

Time reclaimed

Another of Paul Virilio's contentions in the 1980s was that the control of time was contested, as a 'chrono-politics'. Unions could deploy the industrial strike, for example, to force a break in the machine time of industry; a general strike is a 'barricade in time'.[68] Over many decades of struggle, organised labour won working conditions that were measured in time, principally the 35-hour week, guaranteed annual holidays and other forms of leave. Labour time remains a key component in the work process: all workers, whether casual, part-time or full-time, are paid by the hour – but many now work more hours than they are paid for.

The breeching of work's spatial boundaries by mobile networking has unravelled time limitations on work, such as the 35-hour week. The obligation to continue working at home – on laptops, tablets or smartphones – has expanded the time committed to work, in many cases beyond 50 hours per week. The *Squeezed for Time* report on Australian work conditions found in 2012 that 28% of surveyed men work 48 hours per week or more; another finding was that 72% of men working long hours would prefer to work at least half a day less.[69] The decline in influence of organised labour, as union membership decreases, has coincided with the general use of portable networked devices for work purposes in leisure hours. Many of the devices themselves, and the personal organising apps designed for them, have as their goal the managing of information and messaging, with the assumption that 'being managerial is our highest goal', as Rebecca Solnit remarks.[70] These factors have combined to produce time-stressed individuals burdened with the endless task of sorting, answering and sending messages. Developments such as Google Instant, which returns search results before the user has finished typing the query, are designed to accelerate communication even further, allowing for increased volumes of information to be accessed and absorbed.

One remedy for this saturation of time with messaging is the media fast, in which social media users withdraw from all online activity. Many of Sherry Turkle's young respondents speak with pride of their ability to manage constant connectivity, but others acknowledge anxiety, even despair, at the glut of stimuli demanding their attention. Some describe their self-imposed digital fasts as an act of reclaiming time for activities, such as meeting 'real' friends – in contrast to managing an array of Facebook 'friends'. Others look for 'authenticity' in off-line engagements, a quality perceived to be lacking in social media connections based on performance of the self.[71] One interview subject abandons all digital connection, sacrificing 'three hollow conversations' in favour of 'one really nice social interaction with one person'.[72]

The main motivation for media-fasting, however, is to reclaim time from the barrage of information coursing through mobile networks. Some take a small step, such as replacing their smartphones with 'dumb' or conventional mobile phones. This has the effect of drastically reducing the number of incoming messages. One mobile phone user who took this course of action in 2015 stated that his need to check messages on his smartphone had 'felt like an addiction'. Without it, he is 'not so easily distracted any more'.[73] For those who undertake a complete digital fast, this is usually a short-time deprivation only: a week, perhaps, to 'cleanse' the self of messaging and dependence on apps.

Four Australian members of Generation Y agreed to fast for a week in 2011 as a social experiment, reporting their experience to a journalist. All four recorded an initial anxiety at the prospect of losing their smartphones for a week. One 29-year-old admitted to a feeling of 'dread' when he first walked to work without his mobile phone. The loss of Google Maps left him unable to find city locations. Realising that he would need to resort to 'manual' means of finding his way – asking for directions – he found his faith in 'the goodwill of humans' restored. Later, he came to enjoy the fact that he could not be contacted by phone: a new 'sense of calm washes over me'. The greatest change was at night: he usually checked emails before going to bed, causing sleepless nights thinking about work. Without the phone and the checking of messages, he slept 'like a baby'.

The other three participants in the social experiment reported similar experiences, an initial anxiety replaced by a surprised, even pleasurable, realisation that life can be lived even without smartphones. One 25-year-old student was 'amazed' by the amount of information she took in during lectures without her smartphone tempting her to Twitter and Facebook. A 24-year-old project manager found that her disappearance from the social grid was the most significant outcome of her fast: 'it was a really liberating feeling that no one knew where I was, not even Apple'.[74] All four experienced something of 'the joy of disconnecting', a temporary joy encouraged by various self-help groups and popular movements. All four, however, returned to their smartphones after a week.

4
Space and Displacement in Contemporary Art

One thousand and one Qing dynasty chairs set up in Kassel, Germany, awaiting 1,001 visitors from China. A cargo ship poised mid-journey in the Atlantic. A derelict abandoned factory. A detention centre in the desert. Six unemployed men, standing in a line joined by a collective tattoo. Faces painted on a footpath with water, evaporating in the sun. An arm wrapped in the Stars and Stripes, holding uprooted hair. A lost score broadcast from Auschwitz on a railway platform in 2012. An exhibition in Vienna in 2014 by an Australian artist based on the Japanese 'Capsule Tower' built in 1972. A ruined house reclaimed and populated by artists and the unemployed. These are all images of displacement in recent works of art. Space is rendered here as territorial space on a global scale: the space in which people, information and commodities move.

The idea of displacement is often treated obliquely by contemporary artists. This may involve an emphasis within artworks on the nomadic, or the combination of disparate cultural elements. Other works take a more direct, even blunt, approach in depicting the social dislocation found in many parts of the contemporary world. In this chapter, the theme of displacement is considered in the context of the globalist aspect of contemporary art, itself a reflection of globalisation. The intensified movement of goods, information, capital, images – as well as tourists, refugees and displaced persons – around the world provides the setting for contemporary artists' treatment of space and displacement.

Global art

Contemporary art has increasingly been conceived as global art. Art theorists and art historians searching for a successor term to modernism and postmodernism have proposed, among other terms, network culture and globalism, as the cultural conditions reflected by contemporary art. The critic and theorist Rex Butler has suggested that the 'new style or movement of art that comes after postmodernism' should be called globalism,[1] incorporating both the impact of globalisation on the concerns and content of contemporary art, and the international circuit of major art events at which the most recent artworks are showcased.

The art theorist and curator Nicholas Bourriaud proposes as the successor to postmodernism an 'altermodernity' comprising a 'translation-oriented modernity'.[2] For Bourriaud such a conception of contemporary culture corresponds to the globalised world order, a modernity 'born of global and decentralized negotiations, of multiple discussions among participants from different cultures'.[3] Such a culture must be 'polyglot', because 'the immigrant, the exile, the tourist, and the urban wanderer are the dominant figures of contemporary culture'.[4] Altermodernity for Bourriaud embraces the styles and techniques of modernity as 'one phenomenon among others', to be explored in a 'globalised culture busy with new syntheses'.[5] The global network becomes a space of exchange, of diverse representations of the world, in which translation of ideas and representations plays a crucial role in 'discussions that will give rise to a new common intelligibility'.[6]

Bourriaud cites as exemplary artists in this regard Gabriel Orozco, Thomas Hirschhorn, Jason Rhoades and Francis Alÿs, all of whom express the wandering aspect of modern urban life first defined in Baudelaire's figure of the flâneur. This artist-figure becomes 'one flesh' with 'the multitude, among the ebb and flow of movement, in the midst of the fugitive and the infinite'; such an artist is 'a kaleidoscope gifted with consciousness', in Baudelaire's phrase.[7] The nomadic function celebrated by Baudelaire is intensified in the global age, as artists wander not just through cities but across continents.

Francis Alÿs, who was born in Belgium but lives and works in Mexico, typifies the creative negotiation of geographical displacement endorsed by Bourriaud as a principal feature of contemporary art. As the curator and art theorist Wayne Tunnicliffe has noted, 'moving through cities and between countries and locations has often been the nexus of Alÿs' art'.[8] In some of his artworks, this nomadic aspect, expressed as the act of walking through urban spaces, is documented and becomes the central focus of the work. The 2004 video work *Railings* depicts the artist walking through the streets of London, tapping with a wooden drumstick the front railings of the properties he passes. His path through London is a musical one, with music made 'from the fences that define the boundaries between public and private spaces'.[9] Other Alÿs performance events, documented in video, photography and other media, have involved him pushing a block of ice through the streets of Mexico until it melted (*Paradox of Praxis 1*, 1997), and enlisting 500 volunteers in Peru to move a sand dune a few inches by shovel (*When Faith Moves Mountains*, 2002). Alÿs regards these arts events as 'interventions', fragments 'of the city that I am mapping'. These interventions are often inspired by the process of walking through a city: 'As an artist, my position is akin to that of a passer-by constantly trying to situate myself in a moving environment.'[10]

The role of translation assumes a central importance for Bourriaud's envisaged cultural condition of an 'altermodernity', succeeding that of postmodern identity politics and the postcolonial emphasis on alterity. He argues that 'one might define contemporary art in terms of a criterion of translatability, that is, according to the nature of the contents it transcodes ... and inserts ... into a signifying chain'.[11] The various instances of transfer and translation practised in contemporary art constitute 'a practice of displacement', involving 'the passage of signs from one format to another'.[12] Bourriaud offers as examples of this practice the work of Korean artist Kim Soo-ja, which combines a Tao-inspired vision with minimal art and ancestral motifs; the Thai artist Surasi Kusolwong fusing Thai folk and popular culture with minimal and conceptual art; and Indian artist Navin Rawanchakai installing the aesthetic of the Bollywood movie poster alongside Hollywood science fiction within a narrative epic in the style of conceptual art. These works have in common a focus on translation: 'elements belonging to a local visual or philosophical culture are transferred from a traditional universe in which they are strictly codified and fixed to one in which they are set in motion and placed beneath the gaze of a critical reading'.[13] Displacement of cultural material, from one location or context to another, is thus a central component of Bourriaud's analysis of the contemporary cultural condition.

The global network of contemporary art

The art milieu has become, especially over the last decade, a global network. Apart from the exhibitions held in the major museums and art galleries of urban centres, the international biennales, triennales and other recurring shows have established themselves as defining events of contemporary art. Of the world's art biennales, Venice, Sao Paolo and Sydney are the oldest, while many other world cities have expanded the circuit of international art events by founding biennales. One feature of globalist art, as Butler has observed, is that it is 'defined not by artists and artworks but by curators and exhibitions'.[14] Large-scale exhibitions such as Documenta or the Venice Biennale are conceptualised by curators along loosely binding themes, with international artists chosen to respond to these themes with new works, or with relevant existing works.

The curatorial concepts are often politically urgent and informed by critical theory, political theory, radical continental philosophy and other academic disciplines including cultural geography and anthropology, as well as art history and theory. As an example, Documenta – the world's premier art event, held every five years in Kassel, Germany – was curated in 2007 by Roger Buergel and Ruth Noack, who posed three questions to which artists were invited to respond: Is Modernity our antiquity? What is bare life? What is to be done? The second two questions, inspired by Giorgio Agamben and Lenin, provoked responses in photographic and video works documenting refugees, detention camps, the aftermath of war and social inequality. The 2012 Documenta, curated by Carolyn Christov-Bakargiev, eschewed an explicit theme, but was organised around a set of ideas, crystallised by the curator as 'a significance given ... to the act of being emplaced'. This notion of emplacement entailed 'the conditions in which artists and thinkers find themselves acting in the present – of being "on stage"/"under siege"/"in a state of hope"/"on retreat" '.[15] Many of the recent artworks shown at biennales and other global art events favour digital photography and video, as well as other forms including installation and performance, in documenting social and political realities. The idea of place is often foregrounded in these works. Christov-Bakargiev articulated the vision of Documenta (13) in 2012 in these terms, whereby artists often engaged with the various sites of the exhibition in Kassel, Kabul, Alexandria–Cairo and Banff, 'unfreezing the associations that are typically made with those places and stressing their continual shifting'.[16]

The best international art transcends mere propaganda, or the meagre illustration of theory. Artists are explorers within the networks of global images. Using the word coined by Nicolas Bourriaud, the artist is a 'semionaut', navigating the virtual seas of global images, producing 'original pathways through signs'.[17] This is an updating of Marshall McLuhan's description in 1964 of artists as the antennae of their society, in that they pick up conceptual shifts and investigate them in creative works. McLuhan's formulation of the artist was itself borrowed from an assertion made in 1934 by the poet Ezra Pound, that 'artists are the antennae of the race'.[18] McLuhan added the metaphor of radar to that of the antenna, construing art as 'an early alarm system... enabling us to discover social and psychic targets in lots of time to prepare to cope with them'.[19] The most striking artworks probe the fault-lines and fissures of contemporary meaning. They play with contradiction, or expose the processes of meaning-making, national identity formation, wilful acts of cultural forgetting or the consequences of migration or forced displacement. The best artworks evoke ambiguity of meaning while resonating with emotional power. They may be intellectual feats and visceral emotional statements at the same time.

Images of displacement

Because contemporary artist-semionauts are operating within a globalised regime, many current works have a geopolitical aspect. Contemporary art should be understood in relation to the widespread perception that humans have altered their orientation in space-time. The previous orientation in space and time has been 'dis-placed', it is widely assumed, by 'the great liberal utopia of globalisation, an ordering of the earth and its beings which claims to do away with our orientation'.[20] The processes of globalisation – commonly understood as an intensified flow of capital, information and people – function on many levels. Migration, within and across national borders, is a significant factor, not least with regard to the trauma of displacement. Digitised images are also part of this global flow, as is the capital invested in the global financial system and international art market. As Mitchell Dean observes, 'the rationality of globalisation portends an overcoming of space and time in a great frictionless circulation of things, information and people'.[21] Much of recent art has been concerned with refuting this liberal-utopian vision, or at least with exposing some of the friction within the mechanism of world order.

Sound artists have played with the disorientation, or spatial re-orientation, provoked by global connection and movement. The American sound artist Bill Fontana, for example, has built 'harmonic bridges' between cities, including San Francisco, Cologne, Kyoto and Sydney. Fontana uses recordings of the urban environment as the basis for musical information. He has written of his sonic displacements: 'the relocation of ambient sounds to urban public spaces is a radical attempt to redefine the meaning of the acoustical context in which the sound sculpture is experienced'.[22]

Fontana's relocation of sounds has involved the 'acoustic haunting' of ruined urban sites by the audio of the present, as achieved in his work *Distant Trains* (1984), in which the sounds of Germany's busiest train station, Cologne, were played at the site of a ruined former station in Berlin. Other works displace the sounds of natural environments into urban spaces. Fontana is best known, however, for his 'satellite ear bridges', in which the live sounds of one metropolis are broadcast in another, far distant city. In one work, he recorded the urban soundscape of San Francisco, including its harbour, and broadcast this aural cityscape over loudspeakers in the city of Sydney. The sound-map of one harbour city was superimposed on another, creating a confusing, yet agreeable, aural displacement and layering. In other works, such as the ear bridge Cologne–Kyoto (1993), the ambient sounds of the two urban environments are mixed and broadcast to both cities, allowing an audio fusion of the two cities to emerge from the sonic displacement.

More typically, however, contemporary artists express ideas of displacement through visual images. Objects and places are given visual representation and left to resonate within circuits of knowledge and patterns of local and global orientation.

The photographic work *Panorama, Mid-Atlantic* (1993) by American artist Allan Sekula focuses on the objects and movement of global trade. The containers on the ship deck in Sekula's *Panorama* are densely packed, forming a geometrical barrier to the horizon's natural play of light. We do not know the nature of the freight within the shipping containers; we do not know the destination of the ship, except that it is poised mid-Atlantic. Sekula has frozen this cargo midway on its journey. He has exposed an aspect of global trade that is normally kept hidden from view: the long, unglamorous journey of goods from point of manufacture to markets where they will be bought and consumed.

Sekula's documentary photographic practice pursued a Marxist aesthetic strategy. His critical realism was a kind of 'making-visible', an attempt to 'make apparent the underlying forces of society, the invisible base that allows its visible superstructure'.[23] In his many works on international marine transport, Sekula aimed to disclose the unseen labour, organisation, apparatus and processes that work, behind the scenes, to provide our consumer goods. He demystified the modern secular magic of the global commodity. We also glean from this image that beneath the constant whirl of commodities, signs, images and information, is the much slower, often arduous, passage of international trade.

Artists are often fascinated by the underside of globalisation – by the failures, the victims, the unseen casualties of the shining new networks of communication and trade. In the video work *Bade Area* (2005), Taiwanese artist Chen Chieh-jen explores an abandoned factory site. This derelict space is typical of industrial sites that once represented Taiwan's 'economic miracle', now forgotten since the departure of industry to new territories offering cheaper production and labour costs. This ruthless economic logic, central to the globalised economy, has left the artist 'enraged'.[24] For the *Bade Area* video work, he asked laid-off employees to revisit their former workplace – but in this context the ex-workers are adrift like ghosts haunting their former abode. They are displaced in both time and space, 'like spectres in mourning, reminiscing their past life but lost in the present, not knowing where their future lies' (see Figure 4.1).[25]

In the office, a globe on the desk is broken, while a map of the world is torn and frayed. There is a still-connected but ancient computer. These symbols of a previous information system and world order consign the factory to derelict, pitiful status. Elsewhere in the video, workers attempt to set up a karaoke machine on a roof, while

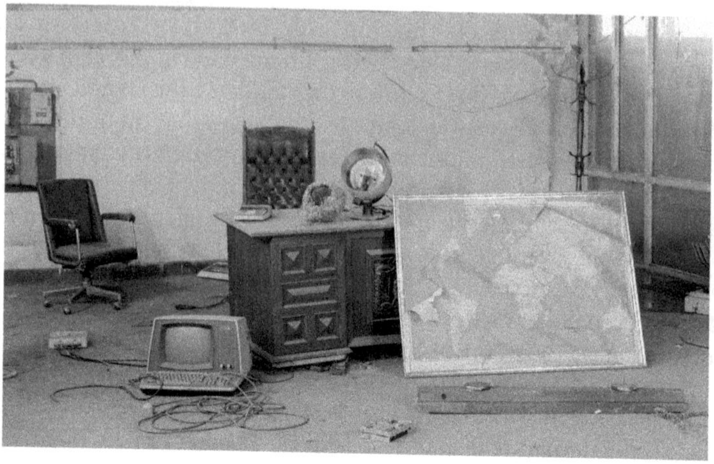

Figure 4.1 Chen Chieh-jen, *Bade Area*, 2005 (production still). Courtesy of the artist and Main Trend Gallery, Taipei

a battered sign heralds 'The Majestic Town'. The Bade area is revealed as a non-place, an ex-place. It offers a 'parable of globalisation',[26] standing for all the sites of production that have been superseded and abandoned, that have had to make way, that have been left to crumble as unloved and forgotten former zones of activity.

Another work based on the ruins of a previously functioning architecture was exhibited in Vienna in 2014. *Never Forever*, by the Australian artist Jai McKenzie, was inspired by the artist's visit to the Nakagin Capsule Tower in Tokyo. This building had been designed by the Metabolist architect Kisho Kurokawa, and was built in 30 days in 1972. The Capsule Tower, built according to Metabolist principles, was composed of 140 inhabitable pods, intended to be removed, re-arranged and eventually destroyed. By 2014, the pods were mostly unoccupied and were scheduled for demolition. McKenzie's photographic work depicts the Capsule Tower as a decayed building and failed architectural project; her exhibition displaces the building in time and space, sifting through the architectural remains, compiling a 'rearrangement of objects repositioned within time, space and imagination'.[27] The Japanese Metabolist architectural vision, which fused urban design with ideas of organic biological growth, is re-imagined in this work as a meditation on time, impermanence and ruin.

A fairy tale of displacement

At the 2007 Documenta exhibition, Chinese artist Ai Wei Wei offered an ingenious comment on the theme of mass migration in the globalised age. His work *Fairytale* involved the transport of 1,001 Chinese citizens to Documenta in five stages. Awaiting their arrival were 1,001 wooden chairs from the Qing dynasty, spread throughout the four main venues of Documenta. These antique chairs attested both to the Chinese past and to the contemporary Chinese citizens they would serve. The artwork to a large extent comprised the act of migration – the feat of physically moving 1,001 Chinese citizens, who had not previously left China – to Kassel in Germany.

This conceptual work was at once a global fable – the title *Fairytale* referred to the Brothers Grimm, former residents of Kassel and to the 1,001 nights of the Scheherazade story – and a provocative act in the context of the Chinese Communist Party's sensitivity to travel and international relations. The art theorist Terry Smith considers *Fairytale* 'a delicious parody of the restrictions to travel that characterised China during the Mao years'.[28] The welcoming chairs, arranged in clusters, fostered informal and lively discussions amongst the Chinese once they arrived, enabling expression of 'civil society in microcosmic form – exactly the kind of formation most threatening to Communist Party centralists'.[29]

Ai Wei Wei has been able to exploit his standing as a major international artist while maintaining a dissident stance within China. *Fairytale* exposed the contradictions in the party's partial embrace of globalisation and the market while attempting to retain authoritarian control over its citizens. The 1,001 Chinese participants in *Fairytale* experienced life outside China for the first time: for one month they were free to engage in the cultural and political pursuits available in a European city. Within China, however, such freedoms may be abruptly curtailed. Despite his international status, Ai's role as figurehead of dissent has earned him severe attention from party officials: in 2011 he was detained by authorities for 88 days.

Ai Wei Wei is also an outspoken critic of the remorseless surge of modernisation undertaken by the Chinese Communist Party. Progress is pursued in China today as pitilessly and unquestioningly as it was advocated in the West by Henry Ford and F. T. Marinetti in the early years of the twentieth century. Another work by Ai Wei Wei at Documenta 2007 was a huge sculpture made from discarded

Chinese antique doors. Entitled *Template*, this gigantic work opens to the past. It evokes the discarded past in its very substance. *Template*, like many of Ai Wei Wei's works, uses as its material the relics of the past jettisoned by the implacable Chinese drive to progress. These materials carry previous time and place into the stuff of the artwork. The sculpture is a contemporary form made from temporal displacement. It is a testament to a former time, former places, an earlier culture – all swept aside by the onward march of the state.

Displaced persons

In his book *Homo Sacer: Sovereign Power and Bare Life* (1998), Giorgio Agamben revived the distinction made in classical Greek thought between *zoē* (bare life) and *bios* (life of the *polis* – the political or good life). For Agamben, sovereign power, including that of the modern state, has always entailed the attempt to control and regulate bare life; this biopolitical impulse is nowhere more evident than in those 'states of exception' where sovereign power has been most apparent. In this context, Agamben regards the camp as the 'hidden paradigm of the political space of modernity'.[30] The camp is 'the pure, absolute, and impassable biopolitical space' because it is 'founded solely on the state of exception';[31] it is 'the space that is opened when the state of exception begins to become the rule'.[32] This formulation is true not only of the concentration camps of World War II, but also of the contemporary camps set up for refugees and political prisoners. For Agamben, the camp is 'the hidden matrix and *nomos* of the political space in which we are still living'.[33]

Agamben's treatise on bare life and the figure of the camp as the 'biopolitical paradigm of the modern'[34] have been extremely influential in critical thought and curatorial practice. Agamben has built on the theorising of the biopolitical – the intersection of power and the body – developed by Michel Foucault, fusing it with a discourse on political theory and the state inaugurated by Carl Schmitt, Hannah Arendt and other theorists. Agamben's conception of the camp and bare life has gained great purchase in the contemporary context of globalisation, in which the movements of refugees, asylum seekers, prisoners of war and migrant workers are political realities.

The impact of Agamben's thought on contemporary art – as reflected in the enshrining of the question 'What is bare life?' as a central premise of the 2007 Documenta – may be attributed in part to his crystallisation of a complex argument in the concrete figure of the camp. Contemporary images of displaced persons held in prisoner of war or refugee camps are familiar to a mass audience due to their circulation in the media. It is readily apparent that sovereign states, including liberal democracies, operate spatial exclusion and enclosure such as manifested in the detention camps for refugees, or those for 'unlawful enemy combatants' held at Guantánamo Bay and other locations. As Mitchell Dean has remarked, these camps represent the 'spatialisation of the state of exception', as a place where anything is possible, 'beyond the language of law and right, a kind of territorialisation beyond law, an orientation without order'.[35]

Agamben has provided a critical reading of globalisation, focusing on the techniques 'designed to impose non-localised forms of order on specific locations in the world, as and when needed'.[36] The camp is for Agamben the emblem of this 'dislocating localization', which embodies displacement and disjunction in its functions. The camp, and its contemporary 'metamorphoses' such as waiting areas at airports, constitutes 'the fourth, inseparable element that has now added itself to – and so broken – the old trinity composed of the state, the nation (birth), and land'.[37]

The theme emanating from many contemporary artworks, including those exhibited at recent Documentas and Venice Biennales is violence: of war, occupation, displacement, hatred. This violence is seen to emanate from and across states unable to control internal hostilities, unable to deal with the flow of refugees, immigrants or victims of war across their borders. A related theme concerns the barriers erected in hostile and fearful times: at detention camps and refugee camps, within brutalised or paranoid states. Detention camps erected in liberal-democratic states, troubled by the influx of refugees, also feature in contemporary artworks.

Australian artist Rosemary Laing's large colour photograph of the Woomera detention centre in Australia is a stark testament to the force of barriers. The title of the work, *Welcome to Australia* (2004) is brutally ironic – but the physical reality of the detention centre in the desert is still more brutal. Laing has photographed the exterior of the camp as a severe construction of barbed wire fencing and search lights. It is a zone of exclusion mounted in the desert, with no sign of humanity or any living thing. Erected during the years of the conservative Howard government, these detention centres were devised to accommodate refugees and asylum seekers while they were – to use the terminology of the official government rhetoric – 'processed'. These detention centres exemplify the political state of exception constituted by the camp, as articulated by Agamben. The goal of these buildings, like the aim of the Government directives that enabled them, is dehumanisation. The asylum seekers are to be locked up and kept out of view of the Australian populace, while they are 'processed' – like objects or pieces of paper – by the government bureaucracy.

Laing has conveyed the iron will of the government's refugee policy through this stark image of its creation in the desert. This is the face of Fortress Australia. The forbidding boundary fence, the barbed wire, the searchlights, the prison-like buildings, the harsh surrounding landscape, all conspire to send the message: 'Keep out'. The Howard government gained great political capital with its unyielding immigration policy, playing to fears of refugee hordes and terrorists in sections of the electorate. But this political advantage could only be sustained by a constant repression. The faces of asylum seekers were resolutely kept from view, lest a glimmer of their humanity was glimpsed by the Australian people. Hence the camp is only visible to observers from the outside: it is protected and defined by its exclusionary walls. The dehumanisation of refugees was enforced behind these fortress walls, in this modern-day camp and state of exception, deep in the interior heartland of a wealthy liberal-democratic nation.

Another Australian artist, Mike Parr, has confronted the political issue of refugees and camps in a series of performance artworks. These works, including *Close the Concentration Camps* (2002), *Aussie, Aussie, Aussie, Oi, Oi, Oi (Democratic Torture)* (2003) and *Kingdom Come And/or Punch Holes in the Body Politic* (2005), involved physical ordeal and the endurance of pain on the part of the artist. In *Close the Concentration Camps* and *Aussie, Aussie, Aussie,* Parr's face was sewn with stitches through the skin and lips. In *Kingdom Come,* Parr, dressed in an orange suit resembling those worn by the inmates of Guantánamo Bay, received an electric shock every time a member of the audience ventured past a point marked by sensors.

These performance works function on several levels, beyond a simple political protest against government policies concerning asylum seekers and the 'war on terror'. As the performance art theorist

Edward Scheer has remarked, Parr's acts of physical ordeal, including violence enacted on his own body, were an attempt to embody the brutality of those policies condoning torture and the detention of displaced persons. This brutality had produced 'a similarly monstrous distortion in the image of Australian identity'; Parr's pain-riddled performances were 'a symbolic response to a reconfiguring of the symbolic order in Australia towards violence and paternalism'.[38]

A more subtle piece on the effect of political violence is the video installation *Proyecto para un Memorial* (2007) by Colombian artist Oscar Munoz, which has on its five screens drawings of faces being traced with a brush then slowly fading. It seems that digital manipulation must be involved, until the viewer realises that the faces are painted with water on a footpath: they vanish through evaporation. The work is a simple and elegant evocation of the 'disappeared' in Latin America.

The American artist and urban activist Theaster Gates has concerned himself with urban regeneration, as in his conversion of an abandoned property in a run-down Chicago neighbourhood into a multi-purpose cultural centre. At Documenta (13) in 2012, Gates performed a similar act of urban reconstruction by repairing the ruined building known as Huguenot House in Kassel, originally built in 1826 and badly damaged in World War II. The renovated building became a venue for arts performances, talks and installations, but also a home for marginalised artists and the marginally employed. This act of urban renewal became both a site for art and an artwork itself at Documenta. The aim of its construction was 'to push the boundaries of labor and production to create space for others'.[39] Visitors to the house could witness its inhabitants discussing art, politics and culture at the dinner table. The Huguenot House, named in honour of religious exiles, became, in the twenty-first century, a refuge for another generation of displaced individuals.

The Spanish artist Santiago Sierra has concerned himself, in a series of works, with the 'bare life' experienced by individuals within the darker recesses of globalisation. In the work *250 cm Line Tattooed on 6 Paid People* (1999), Sierra paid six unemployed men from Havana US $30 each to stand in line with a collective line tattooed across their backs. In another version of this work the following year, four Brazilian prostitutes were paid in heroin for agreeing to the inscription of the tattoo across their backs. In other works, Chechen refugees consented to remain sealed inside cardboard boxes for the duration of an exhibition, and workers were paid low wages to perform meaningless tasks such as moving a slab of concrete back and forth.

Sierra's art practice foregrounds those individuals normally invisible in the global economy: the underclass, the underprivileged, the disenfranchised, the lowest paid menial workers, prostitutes, the unemployed, refugees. These individuals are the raw material of Sierra's conceptual artworks; they also represent the economic and political realities addressed by these works. Sierra's intention is to 'draw attention to racial and economic inequities by replaying them' in his artworks.[40] Sierra highlights the degradation experienced by labourers or the underclass by incorporating their indignity into his art: the unemployed or lowly paid receive small amounts of remuneration to perform pointless or insulting tasks; refugees – often the unwanted, unseen or illegal workers in the globalised economy – are paid to hide inside cardboard boxes.

The global reach of the capitalist system, with its manifold inequities and exploitations, is Sierra's primary target; yet he himself has frequently been accused of exploiting the poor and underprivileged in his artworks. Sierra's response to such accusations has been to focus on the contradictions and hypocrisies within the art world. Like the globalised economic order of production, the art world is an international enterprise involving exploitation and inequity in its production system. It is enjoyed in developed nations by affluent individuals whose liberal or progressive attitudes extend only to a certain point. Sierra's interventions into the processes of the global economy also aim to agitate the complacency of the art world. His role, he has stated, is 'to press my finger on sore places and create uncomfortable situations for people who want to have fun in the gallery'.[41]

Colonial displacement

The violence of displacement has been all too apparent, centuries before the current movements across the globe, to indigenous peoples. *Torch*, by the indigenous Canadian artist Rebecca Belmore, addresses the implications of colonialism as forced displacement of indigenous peoples. The arm of Liberty, bound by the Stars and

Stripes, is inverted and holds uprooted long black hair. This image connects liberty with discipline in the inverted bound arm, and 'replaces the torch of enlightenment with the darkness of uprooted hair'. The Age of Reason oversaw the appropriation of land, the violent dispossession or uprooting of indigenes, and their 'binding within the nation-state and its very symbols of liberty'.[42] Colonialism meant uprooting in many dimensions at once: physical, emotional, spiritual, political.

Tracey Moffatt, a member of the Aboriginal 'Stolen Generations' – those children who were affected by Australian government policy to remove certain Aboriginal children from their parents and place them in state care – has played out narratives of dislocation and identity within her art. In the short film *Night Cries* (1990), an Aboriginal woman cares for her dying invalid white mother. Moffatt, who was raised by white parents after being taken from her Aboriginal mother by state authorities, treats the relationship with a degree of ambivalence. Feelings of resentment and guilt are overlaid upon the mixed-race parental relationship. The heightened artifice of the studio setting, with its lurid colours and non-naturalistic background, invites a symbolic reading – or readings – of the unfolding narrative. The death throes of the mother provoke both grief and liberation for the distressed child, who has assumed the role of carer.

Yet Moffatt, who rejects the nomination of 'Aboriginal artist' because of the strait-jacketing connotations and expectations imposed by such a term, produces work of rich complexity, art that resists simplistic post-colonial interpretation. The daughter in this film displays love and obligation as well as resentment towards her mother. A child of the stolen generation of necessity experiences a displaced emotional bond with substitute parents approved by the state. Yet this displaced bond is also a genuine emotional relationship. Moffatt's series of digital images, shot, like *Night Cries*, to highlight artifice rather than naturalism, present complex layers of race, identity, emotional attachment and shards of story-telling. The series *Something More* (1989), for example, depicts broken elements of an outback story involving mixed race, sex, violence, road journey and escape. The narrative suggested by these pictorial shards is enigmatic, yet clearly includes a dynamic element of movement across the land, alongside various racial and gender elements of identity cross-fertilised in unspecified combinations.

Space, place and sound

Several artists exhibiting at the 2012 Documenta used sound to explore the specifics of place. Place is distinguished from space by its subjective nature: place is the subjective rendering of space, the personal appreciation of a section of space or territory. Sound artists and theorists have adopted the distinction between place and space from architectural discourse, which proposed it in the 1960s and 1970s. As architect and theorist Colin Ripley has remarked, place exists in architectural thought as an 'antidote' to the modern idea of space. Space in modernity was an abstract idea, 'a system within which any particular spot could be designated and located'.[43] Modernist architecture constructed a 'homogeneous and abstract built world' which, to architects and theorists coming after the modernist period, could appear disenchanted, 'devoid of poetry' and 'separated from a real connection to our existence on the earth'.[44] This is the context for the emergence of the idea of place as antidote: a more sensitive architectural practice aims to 'strengthen or emphasise the character of a place', enabling a 'poetical dwelling'.[45]

The academic discipline of human geography emerged in the 1970s with a theoretical emphasis on place, as reflected in the 1977 book by Yi-Fu Tuan, *Space and Place: The Perspective of Experience*. Tuan's focus on the significance of human experience in constructing and defining spaces allowed him to state that 'place is security, space is freedom'.[46] Recent writing in the field of urban geography has concentrated on place as the personal, even arbitrary, rendering of urban space. Lucy Lippard, in her book *The Lure of the Local*, defined a city not as a conglomeration of streets and buildings but as 'a layered location replete with human histories and memories'.[47] Rebecca Solnit's innovative work *Infinite City: A San Francisco Atlas* proposed an infinite number of subjective maps of a city, comprising the personal experience of all those who traverse the city's space. For Solnit, places are 'stable locations with unstable converging forces that cannot be delineated either by fences on the ground... or by the perimeter of the map'.[48] A city is construed as 'many worlds in one place', comprising the individual 'atlases' of each inhabitant and visitor, so that 'every place is if not infinite then practically inexhaustible'.[49] Iain Sinclair's *London Orbital* achieved a similar feat in London, re-interpreting familiar urban topography through walking the orbital freeway, a contemporary act of psycho-geography inscribing the city with a subjective vision.[50]

Sound has been deployed by artists to evoke and provoke notions of place, as the sound artist Ros Bandt has observed: 'place is constructed, remembered, embodied, restored and re-created through certain aural signatures that enable us to interact with that place in new ways'.[51] Because sound fills space, it is particularly effective in conjuring the experience of place or 'soundscape', as sound artist R. Murray Schafer articulated in his influential 1977 book *The Tuning of the World*.[52] Generations of sound artists since the 1970s have evoked place using recording technology or invented sound work; this evocation may take the form of a recreation of the sounds within a specific space, or a creative response to the 'sonic habitat' of a specific environment.

At Documenta in 2012, works by Susan Philipsz and Janet Cardiff and George Bures Miller pursued the idea of 'emplacement' through the medium of sound. The Canadian artists Cardiff and Bures Miller have made numerous works that 'transform our perception of space through the innovative use of sound and image technologies'.[53] Their work at Documenta (13), *FOREST (for a thousand years)*, was an audio installation in a densely wooded park in Kassel. Audience members sat in forest surrounded by an array of more than 30 loudspeakers, creating a spherical sound field. A specific place within the park was thus circumscribed by sound. The sounds emanating from these speakers evoked different periods of time through voices and sound recordings. The audience had the sense of experiencing the passing of a millennium from the perspective of this one patch of territory in the park. At one time listeners heard the clanging of metal nearby and at a distance, suggesting perhaps a sword fight; at another time the forest was full of the sounds of crashing trees and falling bombs: World War II. Other moments were quieter and more intimate, including the sound of a mother talking to her child. By employing sound in ways both dramatic and poetic, the artists in this work complicated the experience of place by invoking the passing of time. This is a work not so much of displacement as of the jostling of place by sound bearing the impression of passing time. The sense of place is complicated by the layering on of duration, as conveyed by these sound portraits in time.

The Scottish artist Susan Philipsz achieved a similar feat of inscribing a specific place in Kassel with layers of time evoked by sound. Philipsz' timescale is much briefer than Cardiff and Bures,' however, focusing on a specific period – 1941–1944 – within World War II. The place at which these sounds were focused was Kassel's former Hauptbahnhof, still in marginal use and accessible to the listening public. At the far end of one platform, seven loudspeakers were arranged in a semicircle above the tracks. These speakers played *Study for Strings* (1943) by the composer Pavel Haas, who died in Auschwitz in 1944. Haas, along with the Jewish families of the Kassel district, were deported by train to the concentration camps from this very platform. The score of *Study for Strings* was lost in Terezin concentration camp, but was later reconstructed, and played through seven channels at the end of the platform. The effect achieved by this work was of the past speaking to the present at this haunted place. The music, re-assembled decades after its composition, was in turn fragmented in space as it played through the seven loudspeakers. The music ghosted into the platform from seven different positions, as if drifting in from the past. The Nazi past associated with this railway station was invoked (nearby in the 1940s was one of the most important arms manufacturers for the German war effort). More poetically, the lost music of a Jewish composer was returned to the public, infusing this venue with a melancholy weight of its past. Displacement occurred at this venue through the intersection of place with the plane of time.

5
The Big Now and the Faraway Then: Present, Past and Future in Contemporary Culture

Temporality may be understood as a box of speeds or gears; each human society possesses such a multi-geared orientation to present, future and past. Just as each civilisation has its own *nomos* or orientation to space, so it can be said to articulate its own *chronos*, or perspective on time. As detailed in Chapter 1, the primary orientation of ancient societies was to the past, in which, it was believed, lay the Golden Age or time of perfection. The present was considered a fallen, deficient age, and the future could only be regarded with hope if – according to the doctrine of the cycle of time – it cycled back to the glorious past. Contemporary Western society observes a *chronos* diametrically opposed to that of the ancient world. The principal orientation is to the present and the near-future, where consumption is achieved, information is communicated, the new mobile phone, computer or tablet supersedes the old and digital progress is made. The past is largely dismissed, often contemptuously, as irrelevant.

In this chapter, I consider the range of contemporary perspectives on present, past and future. Consumer network society is fixated on the present, with an eye to the near-future, in which the next model is regarded as an incremental improvement on the current model. This is particularly the case regarding the digital and mobile technologies that comprise the prized communications technologies of the present. The past is construed in commodity culture often with contempt: the present evinces an attitude of disdain for the inferior and 'primitive' condition of earlier times. If the past is viewed in favourable terms, it is through the prism of nostalgia, which is used to sell products. Nostalgia functions within a particular emotional-economic circuitry – commodification – within the largely de-historicised condition of consumer network culture.

There are alternative perspectives on the past found in the work of contemporary artists. Here the past is represented not as nostalgia or commodity, or object of contempt, but as part of an ongoing dialogue with the present. At the same time, in the work of Anselm Kiefer, Thomas Demand, Walid Raad and William Kentridge the past has an enigmatic quality, devoid of the sentimentalising aspects of nostalgia, existing in the present in a complex, often unsettling, manner.

The big now of consumer society

The temporal orientation of commodity culture remains as diagnosed by Fredric Jameson in 1983: a 'perpetual present' that 'obliterates traditions'.[1] Consumerism is built on building blocks laid down in industrial modernity: progress, designed obsolescence, dismissal of the past unless rendered as nostalgia, a utopian sheen projected onto technology, commodity and product. Commercial culture fosters a love of the present and the near-future, an allure intensified in the marketing of information technology: each new generation Apple product offers the look of the future, rendering previous products out of date. Consumer culture, in its relentless focus on the immediate present, encourages a cultural amnesia in its citizens, especially the young. This atrophy of memory is to the advantage of the market: it's easier to recycle music and films from the 1970s, for example, if the originals have disappeared from view.

What is true of consumer culture is even more apparent in media culture. Every night commercial television news provides 'a cavalcade of events: isolated from process, removed from precedent, detached from history. Each news story is an instant, a spectacle, with minimal connection to other instants.'[2] Television news – or audio-visual news footage relayed on YouTube and elsewhere on the Web – is a series of dislocated images. As a result, the procession of televised news events makes little sense. International news is rendered as a parade of violent spectacles, with no context or rationale. Bombings and other killings blend with natural disasters, with no historical depth to give them background or meaning. The cumulative effect on the viewer is bewilderment and anxiety; this reaction stimulates the desire to retreat from the world into a more comfortable insularity, or consumption, or nostalgia for a fondly re-imagined version of the past. Commercial television news has no – or exceedingly little – memory, and it does not desire memory. When the televised events of the present have been broadcast, they are rarely remembered, as they are readily replaced by a new barrage of images.

This is the triumph of event as smoke, the 'hasty' time disdained by the great historian Fernand Braudel in his tripartite model of historical time:

> Its delusive smoke fills the minds of its contemporaries, but it does not last, and its flame can scarcely ever be discerned.[3]

Braudel wanted to 'transcend the event' and 'the brief moments of awareness' experienced in the short time span. For him, historical meaning could only be grasped at the level of the *longue durée*, which is 'the source, for each society, of its own internal logic'.[4] Yet sadly today, in the domain of both the general public and cultural theorists, the *longue durée* remains largely ungoogled.

The disappearing past in Web 2.0

The evaporation of the past in consumer culture is intensified by the overwhelming emphasis on the present found in Web 2.0. There is literally no past – no archive – in the photo-messaging application Snapchat. Photos and text are sent to a recipient, but the sender selects a time – between one and ten seconds – after which the message disappears and is deleted from the Snapchat server. Hacks of Snapchat have retrieved messages which had purportedly 'disappeared'; nevertheless, the intention – and the appeal – of Snapchat has been to situate messaging firmly in the present. Messages disappear within ten seconds, to be replaced by new messages – and there is no archive, no store of past messages, to serve as a memory. The past is empty.

The archiving of the internet in general is a perilous and haphazard venture, as the historian Jill Lepore observed in 2015. Many web pages have disappeared, due to their hosting sites having expired, or as victims of 'link rot' or 'content drift', when the page has moved, been over-written or deleted. Lepore cites a 2014 study by Harvard Law School, revealing that more than 70% of URLs in the *Harvard Law*

Review and other journals do not link to originally cited information.[5] Because the World Wide Web was built without a time-based version control (in which all previous versions of a page would be stored), the internet 'started disappearing' soon after it began as a commercial web-based entity in the mid-1990s.[6] For Lepore, the internet is inherently 'ephemeral, unstable, and unreliable'; despite the efforts of organisations such as the Internet Archive, based in San Francisco, the internet itself 'dwells in a never-ending present'.[7]

Social media organisations such as Facebook fund themselves by selling information on their users – their 'likes' and current obsessions – to advertisers, who are interested only in today and tomorrow, not yesterday or any further back. Social media is a ravenous beast hungry for the present in the form of status updates, likes and daily ephemera, all tapped as clues for advertisers to offer users the means of satisfying their current needs. As Evgeny Morozov has observed, the 'very architecture and business models of social networking sites' are oriented to 'what we are doing and thinking about right now'.[8] He notes that Twitter and Facebook are not interested even in what we were doing or thinking five years ago; as a result, our lives are increasingly 'lived in the present, completely detached even from the most recent past'.[9] There is a commercial imperative for this 'fundamentalist preference for the present',[10] as the advertisers who feed off social media aim to satisfy the immediate needs and preferences of users/consumers. The past, including the past of social media users, is irrelevant to the goals of the targeted market research available to advertisers in the form of 'likes' and current interests.

Lev Manovich has made the further observation that the structure of the social media apparatus privileges the now and the about-to-be over the past. Manovich describes the constant feed of information on Twitter, Facebook and other Web 2.0 outlets as a 'data stream', which he contrasts both to the database format established by the 1990s World Wide Web and to the older means of organising information, the narrative.[11] Facebook's news feed and ticker facilities display constant updates of friends' activities, while Twitter provides a continuously changing stream of information-bursts. The user experiences this information not through browsing or searching, as in a database, but as a 'continuous flow of events'.

In social media, the new is constantly displacing the not-so-new, as Manovich remarks: 'new events appearing at the top push the earlier ones from the immediate view'. In this system, 'the most important event is always the one that is about to appear next because it heightens the experience of the "data present"'.[12] The latest information event is the most privileged, as all other events 'immediately become "old news" – still worth reading, but not in the same category'. For Manovich, this data-stream interface is a 'quintessential modern experience ("Make it new"), only intensified and accelerated'.[13] It is the Web 2.0 post-industrial version of the imperative to the new, first articulated in the age of industrial modernity in the early twentieth century. As befitting the information age, the present is continuously updated as current information, consigning the past to the digital trash.

The technocratic ideology today emanating from Silicon Valley – that only information and network technologies offer the hope of changing the world for the better[14] – mirrors the technocratic programme pursued by industrial modernists in the early decades of the twentieth century. In this respect, the industrialist Henry Ford set the tone in 1916, speaking not only for his industrialist peers but also for his post-industrialist successors in the next century. Ford elevated technological progress while denigrating history as 'more or less bunk'. For Ford and his modernist colleagues, the present and the near future were paramount, while history was to be dismissed as stale tradition:

> We don't want tradition. We want to live in the present, and the only history that is worth a tinker's damn is the history we make today.[15]

The gear of the future: Progress

The onward drive to a digital future is generated from the engine-room of modernity itself: the doctrine of progress. This doctrine holds that the present is an improvement on the past, and that the future will be an improvement on the present. As discussed in Chapter 1, progress is a recent idea, emerging only in the eighteenth century. Twenty-first-century progress is a severely curtailed version of twentieth-century industrial progress, when faith in a technological future was at times unlimited. Today progress jostles with fear of the future due to global warming and a widespread disillusionment with the earlier promise of social betterment through industrial technology. Yet progress survives, in a new form, in contemporary network society, speaking the language of the near-future in a specific context. Progress is now focused almost exclusively on information technology, which in part explains the ferocious intensity of claims made for this sector as the driving force of the future.

The language of progress has defined modernity since its origin in the Enlightenment, when progress meant the creation of fairer and more just societies through the application of reason.[16] Enlightenment philosophers were emboldened in the eighteenth-century 'Age of Reason' to compare the achievements of contemporary European civilisation favourably even with the glories of ancient Greece and Rome. It was envisaged that reason would continue to spin 'infinite progress', or that society would constantly improve through the application of reason, reaching a point of perfection at some time in the future, as Condorcet predicted in 1794.[17] This faith in the future, as guided by benevolent progress, constituted a 'changed consciousness of time',[18] as Jurgen Habermas has described it: this new temporal sensibility was the spirit of modernity, excited by the new and ever-confident in the future. With the advent of the Industrial Revolution, progress became aligned with technological advancement, especially during the Victorian period, when new railway stations were hailed as cathedrals of progress, and technologies of transport, communication and industry were celebrated as the catalysts of an 'age of improvement'.

The fusion of progress with technological innovation was most sharply expressed in the twentieth century. In the early decades of that tumultuous century, industrial production, rationalisation, advertising and marketing converged in the creation of the new: in the manufacture, promotion and selling of a new range of commodities, and in the notion of the new itself.

The most vociferous advocate of industrial progress was also the inventor of the word 'futurist': F. T. Marinetti, poet and leader of the Italian Futurist art movement. The self-styled 'caffeine of Europe', Marinetti sought to uncouple the present from the past, instead yoking the present to the future. Progress 'is always right, even when it is wrong', he wrote, 'because it is movement, life, struggle, hope'.[19] Marinetti's manifestos are hymns to progress and to the glories of industrial technology: 'We declare that the world's splendour has been enriched by a new beauty, the beauty of speed... Why should we look behind us... Time and Space died yesterday. We want to destroy the museums, the libraries', he thundered in the founding manifesto of Futurism in 1909.[20] The reference to the death of time and space relates to Einstein's recently published Special Theory of Relativity; the speed Marinetti found so beautiful was the speed of the racing car, locomotive or any industrial technology.

The Futurists sought to jolt drowsy Italy from its agrarian slumber by embracing the industrial manufacture of northern cities such as Milan. Through their art, manifestos and public provocations, they celebrated a delirium of speed, metal and violent force, declaring war on the past and all its relics. Marinetti's exuberant love of technological speed and progress was also a statement of contempt for tradition, which could only impede the new. His fervently articulated position constitutes the extreme stance in favour of progress, of the future elevated over the past.

The faith in technological progress espoused by modernist zealots such as Marinetti was aligned with the culture of industrial capitalism in the first six decades of the twentieth century. The shock of the new in aesthetics ran parallel with the triumph of the new in production, and the thrill of the new in consumption. Industrial capitalism declared war on the past just as ferociously as had Marinetti in his manifestos. Planned obsolescence was a production strategy: why drive last year's car when this year's is so much better? Advertising sold the virtues of the new more effectively than any manifesto. Coffee was replaced by instant coffee (surely even Marinetti, as the caffeine of Europe, may have had his doubts about that innovation); seasonal vegetables were usurped by frozen, packaged replacements, cotton and wool by synthetics. In the name of progress, nature, it seemed, was improved upon by technological means.

The high point of industrial progress was the 1960s, its zenith 1969, when rocket technology reached all the way to the moon, the space age beckoned, energy knew no limits and popular culture brimmed over with optimism. The media sage of the 1960s, Marshall McLuhan, observed the march of progress in the new age of electronic media: television, radio, rock music, advertising, the satellite. The swarming energy, spontaneity and 'electric speed' of the new media were sweeping aside the staid, linear, predictable world of printed media. McLuhan saw the generation gap and counter-culture in media terms: when Bob Dylan sang, in 'Ballad of a Thin Man', 'Something is happening here, but you don't know what it is', McLuhan heard 'you' as the older generation, and 'it' as electric media.[21]

For McLuhan, the fate of old media was to become the content of new media: plays become the content for films; films become the content for television; newspapers become the scripts of radio and television news. This process was inevitable, as new communication technology created the 'global village' of electric media. Anyone lamenting these radical changes – or failing to understand them by focusing on the mere content of media – was guilty of 'the numb stance of the technological idiot'.[22]

McLuhan's uncompromising rhetoric has been absorbed into the corporate language of the internet age; indeed McLuhan has been hailed as the prophet of networked culture.[23] The McLuhanite division of generations along media lines is also evident in contemporary culture, in which the 'digital natives', who have grown up with the internet, mark themselves off from older generations by means of digital technology. Members of Gen Y reject the newspaper in favour of online information; they audio-cocoon themselves in public space with iPods and earphones, preferring the online community of social networking; they scandalise their parents with a brazen disregard for privacy, as manifest on Facebook pages, blogs and reality TV shows.[24] This generation, and succeeding ones, will – according to the imperative of progress – leave old media technologies such as the newspaper and book behind, embracing instead mobile networked media technologies.

This narrative, culminating in the great collective gathering in the cloud – the network of servers housing all information – has a pronounced utopian aspect. The utopian ardour was present in McLuhan's rhapsodising of 1960s youth living 'mythically' under the auspices of electric media, but it can also be traced in earlier visions of techno-utopia. Marinetti dreamed in 1915 of a future city in which electrical energy is transmitted without wires, and 'intelligence finally reigns everywhere... Every intelligence grows lucid'.[25] In the 1930s, H. G. Wells argued for an 'integrated, world-wide information service' that he named the 'World Brain'.[26]

These utopian longings, from within the age of industrial modernity, have been realised to some extent in the post-industrial era, in the form of the internet. Contemporary rhapsodising of the 'collective intelligence', of which enterprises such as Wikipedia are early expressions, has been led by Pierre Levy and many other writers and commentators.[27] But the fundamental difference between today's hymns of progress and those of Marinetti, Wells and McLuhan, is that faith in industrial progress has long since expired. It went into terminal decline around the same time as the Apollo space programme, in the 1970s. Industry became associated with pollution, waste, environmental disaster and finally global warming. Progress has gone so far into reverse in the industrial zone that the future is now viewed not with hope but with trepidation, and industrial pollution plays the role of villain. Recycling, sustainability and alternative energy are all attempts to undo the damage wrought by previous generations' unquestioning faith in technological progress.

This means that the doctrine of progress has now swung fully behind information technology. Progress is now measured in terms of faster, smaller, more flexible. Marinetti's cult of technological speed survives, but the pertinent speeds now are processing speed and speed of connection. The post-industrial world of information is depicted as clean and free of industrialism's sins: e-books and online newspapers save trees. Corporations can pursue the strategies honed in the first decades of industrial modernity – such as planned obsolescence – while trumpeting their credentials as model green corporate citizens. Each new generation computer or smartphone renders its predecessor out-of-date; at the same time, each new model is a step towards the utopian future: wireless, energy-efficient, sustainable, increasingly immaterial.

With so much force – economic, ideological, technological – behind this march of digital progress, old-fashioned technologies such as the book and newspaper are frequently cited as the future's casualties. The CEO of Amazon, Jeff Bezos, has declared 'that the physical book and bookstores are dead'.[28] Their replacement, in the vision of Bezos and his fellow captains of information, is the vessel for digital text: the Kindle, iPad or other digital device. The proponents of a digital tomorrow view the 'legacy' media forms of books and newspapers with disdain, impatient for their demise.

Objects made of paper and ink: your time is up. If you were animals, you'd be put down in acts of mercy. If you were characters in a film, you'd be described as 'washed up', 'has-beens', grizzled, overweight, self-indulgent and far from pretty: much like Orson Welles' character in *Touch of Evil*, fallen in the trash, his 'future all used up'. The print newspaper is down and almost out, printed magazines are decidedly unhealthy. If newsprint is not long for this media-sphere, then the book cannot be far behind it. Why should the plant-matter book survive, when its successor – environmentally friendly, convenient, opening to a vast digital immaterial library – is already here?[29]

Certainly there are alternatives to progress even within contemporary network society. Recycling has emerged as an alternative to planned obsolescence; conservation in all its forms aims to preserve the past rather than supersede it. There are aspects of recycling in popular culture: Simon Reynolds has documented the 'retro-mania' within pop music, pop's 'addiction to its own past'. He cites fashion critics who pinpoint 1965 as the pinnacle of newness and innovation, after which fashion entered its ongoing pattern of recycling and pastiche. Pop music, according to Reynolds, began recycling itself around 1968, and is currently awash with borrowings of past styles. Yet Reynolds also notes that because music history 'is splayed out as an atemporal smorgasbord', the original meaning or context of the music becomes irrelevant, causing 'historical depth to drop out'. Past styles are revisited, that is, 'without much affect', as pastiche or raw material.[30]

There are counter-progress impulses within contemporary culture, as traditional values and techniques are asserted over technological advance and convenience. The International Slow Food Movement, for example, founded in 1988, is opposed to 'Fast Life' as well as fast food; it is constitutionally anti-Marinetti (the Futurist preached against slowness as 'evil' because of its nostalgia and pessimism[31]). In common with the Slow Food emphasis on traditional custom, a minority of young consumers listen to vinyl records, read books and even newspapers, in defiance of the edicts of digital progress and in the spirit of 'the old ways are the best ways'.[32] Yet these remain minority instances. The post-industrial incarnation of the doctrine of progress has allowed the face of the market to remain resolutely set to the future, guided by the imperatives of growth, development and faith in information technology.

The past as villain

Given the orientation to the present in the consumer media world, history is not much welcome in this precinct. Occasionally, however,

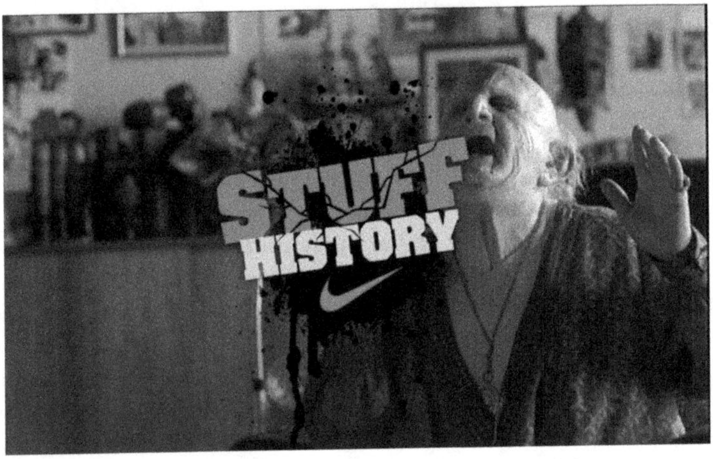

Figure 5.1 Nike 'Stuff History' TV advertisement (still), 2006. Courtesy Publicis Mojo. Actor: Terrance Camilleri

it may be wheeled out and given a starring role, if only as villain. In 2006, for example, Nike attached itself to the football World Cup – the world's biggest cultural event – with a series of advertisements for the Australian market featuring 'History'. In these ads, History was personified for our amusement – and disgust (see Figure 5.1).

As embodied in the Nike television advertisements, History is ugly, unpleasant, cantankerous, foreign (some kind of European), of possible Goblin background, shrunken with age, consumed with memories, living in a dressing gown, consigned to a nursing home. He addresses the footballing New World with contempt, deriding its aspirations. Nietzsche may have had such an ogre in mind when he described, in 'The Uses and Disadvantages of History,' the 'withered and dry' condition of one who is too weighed down with history.[33] Yet even Nietzsche, in that essay of 1873, proposed a balance of active forgetting and remembrance. It is doubtful he would have embraced the abdication of memory and rejection of history found in the modern consumer society. Most definitely, as the Nike ad reports, History is 'against us'. Nike's rejoinder, on behalf of the New World, is to 'stuff History' – a crude reprisal of Henry Ford's equally crude modernist dismissal of history as 'more or less bunk'.

Nostalgia

The past serves a purpose in the consumer world only if mythologised or waxworked as nostalgia, when it can be used to sell something. Nostalgia is a Swiss invention; or at least, the word was coined by a Swiss doctor in the late seventeenth century, from the Greek: *nostos* – 'homecoming' and *algos* – pain or ache. The condition was initially known as *mal du Suisse* or the Swiss illness, due to its frequent appearance in Swiss mercenaries abroad, pining for the mountain landscapes of home. There is even a loose connection between the newly identified disease of nostalgia and the founding of Australia as a European colony: the English botanist Joseph Banks wrote in his journal, on the long sea voyage to Australia in 1770, that the sailors were 'pretty far gone with the longing for home which the Physicians... esteem as a disease under the name of Nostalgia.'[34]

In the nineteenth century, nostalgia lost its status as a disease, instead acquiring a romantic hue as a heightened emotional state, a paradoxical sweet ache or delicious melancholy. Today nostalgia is understood as a personal emotional attachment to the past – or rather to a version of the past. Most individuals experience nostalgia to some degree: a longing for one's past childhood, family home or perhaps country of origin. But when used for commercial purposes, it is an entirely sentimental construction of the past.

Most importantly, it is an extremely lucrative construction of the past within the media industry. An individual may have an emotional attachment to the music of his or her youth and young adulthood, but radio stations capitalise on this charge of affect by programming music solely of this period. This is the domain of the hits and memories radio format, also known as adult classic, original hits and timeless memories, and radio recall. This commercial radio formatting freezes the flow of pop music into a museumised version of the past, selling the loyal listeners to advertisers. This form of commercial nostalgia sells and promotes the past as commodity.

Similarly, period television series such as *Downton Abbey* or *Madmen* construct a glamourised version of a historical period, which can fuel new vogues in retro-fashion or a fondness for earlier trends. The past is rendered as exotic in these period reconstructions, in which production design plays a leading role in evoking the design aesthetic of specific historical periods and cultures. Most viewers, of course, did not experience the 1920s or 1960s the first time around; for them these reconstructions represent nostalgia for a time and place otherwise unknown. This is televisual nostalgia as simulacrum: artfully constructed phantasmogoria of the past, for those who did not experience the original.

The current vogue for televisual nostalgia indicates a fondness for peering back at these past worlds, however idealised and fictionalised they may be in construction. No doubt there are contradictory impulses at play as viewers negotiate the differences between past and present. The freedom of the 1960s, when a three-Martini working lunch functioned as a norm, may be contrasted with the greater restrictions of our more bureaucratic, health-conscious, regulated society. At the same time, our position in the present – morally superior due to the benefit of decades of social progress – is confirmed: we can look back and rebuke the earlier period for its shortcomings – the oppressive class system of early twentieth-century Britain, or the sexism and reprehensible attitudes to women of pre-feminist 1960s America.

In this regard, nostalgia may be construed as the 'ideological twin' of progress, as Christopher Lasch has observed. The veneration of both progress and nostalgia constitutes an 'abdication of memory',[35] in that the past is represented as other, backward, superficially appealing but ultimately undesirable. Nostalgia as refracted through mass media is thus related not only to progress, but to contempt of the past.[36]

'Memory is an artist'

What would an un-nostalgic representation of the past look like? Much recent art is concerned with the event not as an ephemeral moment, but as a political and social act, rooted in time and place, that reverberates far beyond its initial circumstance. Artworks play with contradiction, or expose the processes of meaning-making, national identity formation or wilful acts of cultural forgetting. In nations where events may have been erased from national or collective memory, art may 'intervene to undo this willful forgetting; or the artist may challenge the official record by constructing alternative histories, following a counter-archival impulse'.[37] These practices jostle with official narratives and orthodoxies in what has been described as the politics of memory. Recollection and archive, past, present and memory, may mingle here in a zone where, as the Atlas Group informs us, 'fiction is not necessarily the opposite of non-fiction'.[38]

Many contemporary artists are concerned with the process of memory – but this is memory not as a commodified dose of the past or as sweetened nostalgia. Memory is often revealed as partial, subjective, faulty, contested. The biologist Steven Rose, in *The Making of Memory*, remarks on the imprecise, essentially creative aspects of human memory. Unlike the various forms of artificial memory – especially computer memory – the human mind does not simply invoke information in its act of remembering. Rather,

> each time we remember, we in some senses do work on and transform our memories; they are not simply being called up from store.[39]

For Rose, the subjective, inaccurate nature of human memory contains an element of 're-membering', a variant of 'dismembering'. This is an active process, in which the past is refigured with each act of memory: 'Our memories are recreated each time we remember.'[40]

In recent years, this 'reconstructive' view of memory has gained numerous adherents in the inter-disciplinary field of memory studies.[41] The psychologist and novelist Charles Fernyhough, in his book *Pieces of Light: The New Science of Memory*, defines memory according to the reconstructive approach as: 'a process of constructing something from its parts, in similar but subtly changing ways each time, whenever the occasion arises'.[42] Drawing on Martin Conway's studies of the 'force of coherence' in remembering, whereby memory is made consistent with our current self-hood, including images and beliefs we hold of ourselves, Fernyhough maintains that 'memory is an artist as much as it is a scientist'.[43]

If memory can be regarded as an artist, then artists have responded with an intense concern for memory. The creative re-casting of past events, experiences and processes – which for reconstructive memory researchers is integral to the very nature of human memory – is made evident in many recent artworks. Artists have been concerned to reveal the persistence of the past in the present, in opposition to the veneration of the Big Now in contemporary culture, or the willed forgetting of the past in some national contexts. This fascination with the endurance of the past was also evident in the modernist period, at a time when industrial progress and the celebration of the new in all its forms was weakening respect for tradition. As discussed in Chapter 2, the philosopher Bergson argued that the past 'gnawed into the future' and endured in consciousness. The composer Igor Stravinsky stated in his Harvard lectures in 1939 that:

> Real tradition is not the relic of a past that is irretrievably gone. It is a living force that animates and informs the present.[44]

The novelist William Faulkner made a similar observation in 1950 in his play *Requiem for a Nun*: 'The past is never dead. It's not even past.'[45]

The German artist Anselm Kiefer has carried an obsession with the power of the past into the twenty-first century. His early works, including the series of photographs entitled *Occupations*, in 1969, contained outlawed Nazi references: these works were a deliberate challenge to his German compatriots not to forget the horrors of the Nazi past. At a time when a collective national silence had lowered over the dark period of World War II, Kiefer sought through art to disrupt this national amnesia and confront the past. At the same time, he aimed to reclaim national symbols, such as the forest, from Nazi contamination. Kiefer connects to the mythologies of Celts, Druids and Germanic tribes, with their cyclical conception of time. 'No atom is ever lost', he has stated, which means that the atoms surrounding him and constituting his artworks are 'the tangible remains of former times and long dead people'.[46] The materiality of his works, which can include substances such as straw, ash, clay, shellac and lead, embodies the past in large-scale paintings, sculptures and installations. The process of change can occur in a Kiefer artwork even when it is hanging in a gallery: the straw or other material will age and transform in time; metals may have been dribbled with acid to effect transformation. Kiefer's art is heavy with the weight of the past.

The South African artist William Kentridge mines the concerns of time, memory and the construction of the past in several of his works. His installation *The Refusal of Time*, shown at Documenta in 2012, was a study of the imposition of standardised clock time in the nineteenth century by European colonial powers – and the resistance to this mode of time from within Africa. Referring to the colonial attempts to standardise time across the globe, Kentridge has remarked that 'time is also geography', as the European powers mapped time zones across the world for 'commerce and control'.[47] Kentridge gives voice in this installation work to resistance to the European clock within nineteenth- and early twentieth-century Africa, as the absoluteness of Newtonian time was itself undermined by Einstein's discovery of time's relativity.

In the animated film works for which Kentridge is most well known, he deploys deliberately old-fashioned techniques. His animated films are drawn by hand, in a manner that breaks with the conventions even of old-style hand-drawn animation. Kentridge adds layers of different charcoal drawings onto the same sheet, rather than on separate successive sheets. The result is that each image is composed of traces of previous drawings. The past and the present are compressed into the image by means of the very technique of creating the image. Each moment of the film is a palimpsest. This technique mirrors many of the concerns of Kentridge's art.

The characters in his animated films play out some of the political and emotional struggles within South Africa. His more general theme, however, is erasure. He is concerned with forgetting, both willed and unintentional. His works evoke the suppressed or forgotten, lying beneath the surface of the present. In a note accompanying the work *Felix in Exile*, Kentridge wrote: 'In the same way that there is a human act of dismembering the past, there is a natural process in the terrain through erosion, growth, dilapidation that also seeks to blot out events.'[48] The ambiguity and uncertainties within Kentridge's art are conveyed within this constant engagement or rejection of the past. His technique in the animated films visualises the act of erasure, as the traces of previous frames remain evident to the viewer in the current frame: the past lingers in the present even as it is erased.

In a series of works, the Lebanese-American artist Walid Raad has focused on history as the process of interpreting events. The Atlas Group is a fictional archival organisation, created by Raad. Heavily detailed videos and performance works are Raad's creative version of an archival reckoning of historical process – with specific reference to the 1975–1991 Lebanese civil wars. Raad offers the Atlas

Group project as a vast, all-encompassing archive of these historical events. The officious titles of the Atlas Group documentation connotes ordered objectivity; however, those titles are apt to change their details and dates.

The Atlas Group videos are narrated by a fictional character – yet the archived documents, the events and the city are nevertheless held out to us as real. Raad exposes the difficulty, not with recording these events, but with ordering and defining them – that is, deriving meaning from them. The Atlas Group pronouncements gradually reveal themselves as harbouring lack of clarity and confusion. The Group's obsession with empirical detail, ordering and classification is presented as a parody of the objective mode of 'correct' state or official history. In performances, Raad presents some of the amassed historical detail of the civil wars in a dry archival tone – yet his surety dissolves into the repeated phrase 'It's unclear why...' when more complex issues of cause and effect arise.

The documentation of history as objective record is revealed as impossibly confounded by competing viewpoints, uncertain motivations and the labyrinthine nature of the causal chain. The Atlas Group project, beneath its objectivist surface, is an 'archive of traumatic experience', as Alan Gilbert has observed.[49] Rather than a pristine and objective archive, the project is revealed as a repository of symptoms. The reverberations of a whole series of events are so charged, and so complex in their convoluted sequence, that the empirical record of history must fail to encompass their range and their consequence. Testimony is inadequate. Place and event are displaced within the very telling of their history, and the past remains enigmatic.

The German artist Thomas Demand pursues a singular technique. His works are based either on photographs from media or photographs of the object-world. He builds carefully crafted paper and card sculptures, based on the photographs. The sculptures are then photographed and destroyed. His work thus effects a series of displacements in the process of representation. The final image is a photograph of a model based on a photograph of an object. The final printed image will bear similarities to the original photograph, but the printed image will differ from the original in significant ways: details will be omitted or altered, so that the print is an approximation of the photograph, in the same way that our memories store approximations of remembered sights and events. If one were charged with the task of reconstructing a scene or space from memory, the result would be nothing more than a likeness: recognisable but lacking the details and fine points of an accurate representation. Demand's photographic images share something of this subjective, imprecise aspect of recollection, the re-membering/dismembering process of active memory described by Steven Rose.

Many works by Demand, based on images taken from newspapers, magazines and other media, have probed the slipperiness of memory and the construction of historical truths as forms of myth-making. The 'problematic of commodification' is also visible in Demand's works built on media images, as Tamara Trodd has observed.[50] In contrast to these works, the images in Demand's recent series *The Dailies* are close likenesses of transient scenes, insignificant moments and objects, suspended in time and memory. Each of the dailies is built up from scenes observed and photographed by Demand while walking the streets, each scene then filtered and edited in the painstaking reconstruction process. They are ordinary objects in simple settings, familiar as office, hotel, suburban yard or urban street. At times stark in their reduced simplicity, the images are contemporary still-lifes: banal, forgettable, yet bearing a hermetic appearance. They suggest, as Sophie Forbat has observed, 'the bare compositions of dream, impressions or recollections' (see Figure 5.2).[51]

The 2012 Sydney exhibition *The Dailies* was housed in the Commercial Traveller's Association Hotel, further accentuating the everyday aspect of the images as they may be processed, sorted and recollected in memory. Each of the photographed images was installed in a room in the commercial travellers' hotel, creating the impression that each work represented a brief recollection, a still-life witnessed by the room's inhabitant. As a component of the installation encompassing the cramped hotel room, the printed image suggested a moment captured from the previous day, re-worked in memory, poised in the possibility of narrative (see Figure 5.3).

The title *The Dailies* refers to cinema and the daily rushes viewed in the editing room, selected for inclusion in the greater narrative of the film, or perhaps set aside or cut out altogether. They are glimpses of the everyday, a few of the multitude of images experienced, sorted and edited by the mind 'in the creation of its narratives'.[52] They are

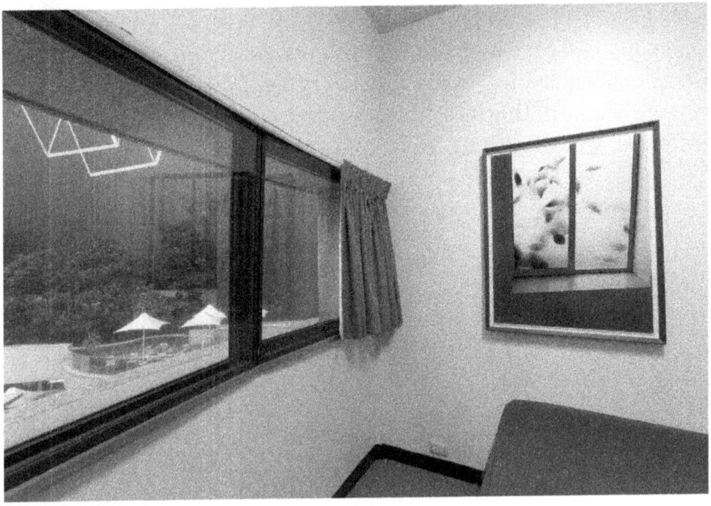

Figure 5.2 Thomas Demand, Installation View: *The Dailies* #14, Kaldor Public Art Project 25, Sydney 2012. Photos: Kaldor Public Art Projects/Paul Green

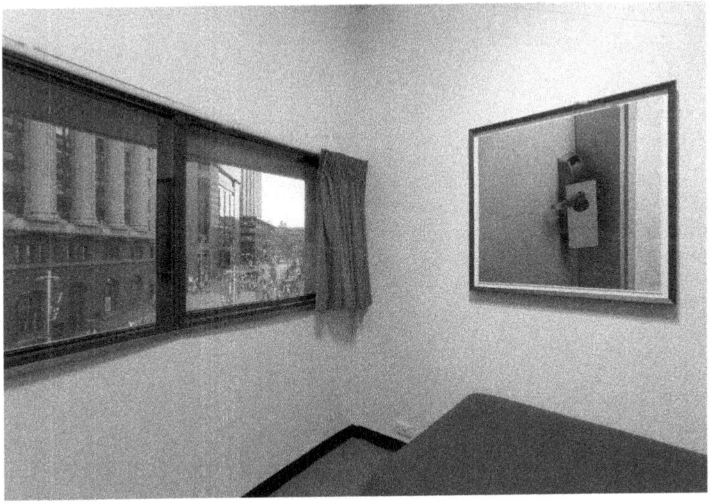

Figure 5.3 Thomas Demand, Installation View: *The Dailies* #12, Kaldor Public Art Project 25, Sydney 2012. Photos: Kaldor Public Art Projects/Paul Green

the dailies of the mind's editing room, also known as memory. *The Dailies*, focused on the ephemeral and the quotidian, shows us a series of enigmatic instants in the process of being sorted. In their rigorous construction, they are representations of memory that are free of sentiment, free of nostalgia and free of contempt.

6
Public Intimacy: The Shrinking Space of Privacy

View from a plinth

Two women occupying a plinth barely big enough for one person, for 11 days, is an act of public intimacy. *Coexisting* was a performance work by the Australian artists Clark Beaumont, one of the *13 Rooms* curated by Klaus Biesenbach and Hans Ulrich Obrist in Sydney in 2013.[1] For the 11 days of the exhibition, Clark Beaumont sat on the tiny plinth, enduring discomfort, squirming for position in the restricted space. A historical reference point for this prolonged act of sitting is the fifth-century Christian ascetic Saint Simeon Stylites, who perched on top of a pillar for many years. A significant difference in this case, however, is that the sitting is done by two people, not one (see Figure 6.1).

As co-inhabitants of one small area, Clark and Beaumont, as individuals, had no choice but to negotiate the space, striving for a comfortable position without discomforting the other. Yet the two performers, who are one artist, are a relationship, and like any relationship theirs will involve compromise, discord, territorial dispute. In living out their relationship on the plinth, they may experience moments of boredom, anguish, irritation, perhaps moments of amusement or even elation. *Coexisting* externalised the inner workings of this relationship, as the 11 day occupation of the plinth was conducted before an audience. Members of the public were free to enter the room at any time, to walk around the plinth and observe the two women on top of it (see Figure 6.2).

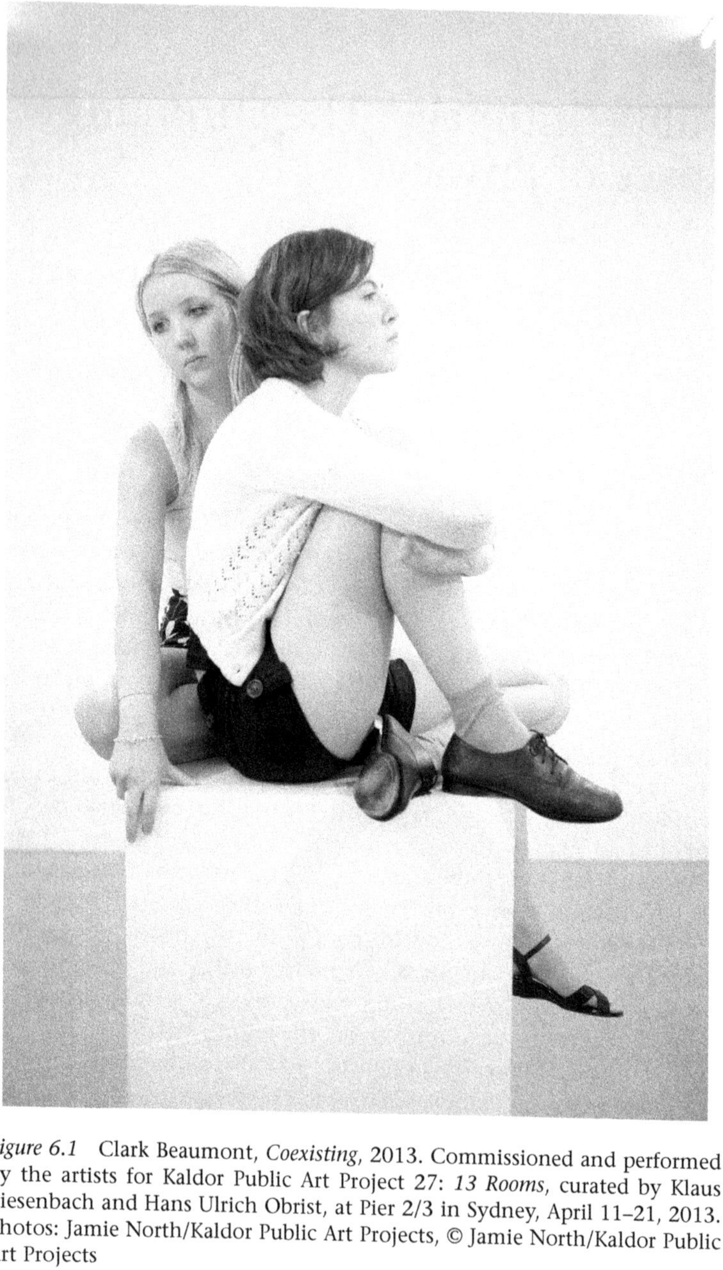

Figure 6.1 Clark Beaumont, *Coexisting*, 2013. Commissioned and performed by the artists for Kaldor Public Art Project 27: *13 Rooms*, curated by Klaus Biesenbach and Hans Ulrich Obrist, at Pier 2/3 in Sydney, April 11–21, 2013. Photos: Jamie North/Kaldor Public Art Projects, © Jamie North/Kaldor Public Art Projects

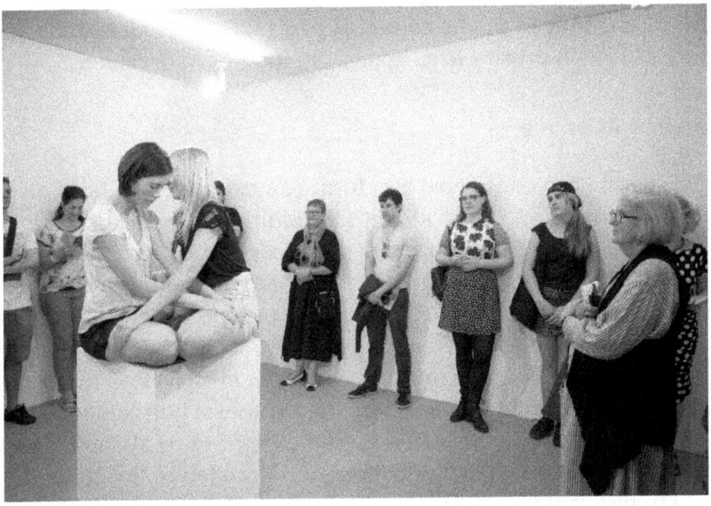

Figure 6.2 Audience views Clark Beaumont, *Coexisting*, 2013. Commissioned and performed by the artists for Kaldor Public Art Project 27: *13 Rooms*, curated by Klaus Biesenbach and Hans Ulrich Obrist, at Pier 2/3 in Sydney, April 11–21, 2013. Photos: Jamie North/Kaldor Public Art Projects, © Jamie North/Kaldor Public Art Projects

The enforced interaction of the artists was physically evident, for anyone in the audience to witness. Clark Beaumont for those 11 days constituted a collaborative self, playing out the dynamics of their two-person relationship in a shrunken public space. This performance chimes perfectly with the recent shift in contemporary culture, in which acts of intimacy are increasingly conducted in public: on social media, on blogs, on reality TV, on mobile phones.

Intimacy is in part a matter of space: the space, or proximity, between two people, and the space – previously private or domestic – in which intimate behaviour is conducted. But intimate feelings and relationships, once confined to the domestic or private sphere, are now performed for everyone to see and hear. Bloggers pour their hearts out, no matter how messily; Facebookers trumpet their intimate encounters; reality TV participants showcase their behaviour in the bedroom, the bathroom and any other room equipped with a camera; the annoying mobile phone user lets everyone else in the train carriage know the details of his latest intimate involvement.

This recent shift in public culture is part of a refiguring of time and space, in which privacy itself appears to have been devalued.

Privacy as space

Privacy has been considered as a form of space. The American attorney and privacy specialist Robert Ellis Smith has defined privacy as 'the desire by each of us for physical space where we can be free of interruption, intrusion' and other unwanted impingements. Ellis Smith's definition continues, embracing an aspect of time as well as space: privacy entails 'the attempt to control the time and manner of disclosures of personal information about ourselves'.[2] Privacy regulation and definition varies throughout the legal jurisdictions of the world, but a common reference is to the 'right to be left alone' definition first articulated by US Justice Louis Brandeis in the 1890s. The right to be left alone encompasses physical or bodily privacy, relating to the space around one's physical self, including the space occupied by one's body. In the age of mass communication, this physical space was augmented by the space of communication, including images and other representations of the self that could be transmitted across space. Territorial privacy encompasses the space of one's home, including yard or grounds: the domestic space.

In the age of the internet, a virtual dimension of space has been added to the conception of privacy: the right to control one's personal information, including the ways in which governments and organisations handle such data. The documents released by Edward Snowden in 2013, revealing the National Security Agency's extensive surveillance of information derived from search engines, social media, phone messages and smartphone apps, alarmed many observers, who had not previously suspected the extent of this data surveillance. This revelation prompted further declarations of the dictum frequently announced in the internet age: privacy is dead. A vice-president at the US information security firm Zscaler decreed in 2014 that 'privacy is dead in the digital world that we live in'. The likelihood that any online information could be easily accessed by government agency or corporation produced the warning: 'unless you are comfortable putting that statement on a billboard in Times Square and having everyone see it, I would not share that information digitally'.[3]

The information privacy component of private space has also generated a new dimension to privacy, at least in Europe: the right to be forgotten. This new plank of European privacy law, which extends into time as well as space, originates from a 2014 ruling of the European Court of Justice that a citizen's fundamental rights could be harmed by information on the internet. Where there is no public interest in publishing this information online, the court held that it is within a citizen's rights to have adverse material removed from the internet. The archival function of the internet – links to earlier stories or information – can from 2014 be redacted according to the wishes of aggrieved individuals in Europe. Google reportedly received 41,000 requests in four days for search results to be removed; it claimed to have removed results 'under data-protection law in Europe'.[4]

The European 'right to be forgotten' legislation highlights differences between Europe and the Anglosphere regarding privacy: the decision was widely criticised in Britain and the US on the grounds that it 'could undermine press freedoms and freedom of speech'.[5] Indeed journalists, who have often expressed frustration at the obstacles to investigative journalism thrown up by defamation law and other media-related legislation, have at times professed impatience with the space of privacy. The British journalist Paul McMullen told the Leveson Inquiry into media practices in 2011 that he believed no one should have privacy. 'Privacy is the space bad people need to do bad things in', he declared. He went further: 'Privacy is evil; it brings out the worst qualities in people.' It is a space used only by evil people: 'nobody else needs it'.[6]

This extreme view is unlikely to find much public support from media proprietors or even other journalists, but it takes its place in a long line of recent public comments either dismissive of privacy as a concept, or disdainful of it. The captains of post-industry have led the charge in this regard, beginning in 1999 when Scott McNealy, co-founder of Sun Microsystems, declared: 'You have zero privacy; get over it.' In 2009, Eric Schmidt, CEO of Google, advised that 'If you have something you don't want anyone to know, maybe you shouldn't be doing it in the first place.'[7] The theme of the 2007 Ars Electronica – an international electronic arts festival – was 'Goodbye Privacy'. Journalists and cultural commentators noticed a shift in the status of privacy, 'from something to be guarded into an anathema to modern living'. This altered attitude was particularly evident on social media, where Facebook has been preaching its mission since 2004: to connect people, to allow them to share, to create an open world. In the light of this vision, privacy concerns began to look like hindrances to connectivity. As a result, by 2007 there were invocations to 'say goodbye to that quaint notion, privacy'.[8]

Facebook founder Mark Zuckerberg was highly visible in the public denigration of privacy. He announced in 2010 that privacy was over-rated, to the point of being outdated. Privacy is 'no longer a social norm', he declared; in its place, Zuckerberg advocated 'radical transparency' through social media. 'People have really gotten comfortable not only sharing more information and different kinds, but more openly and with more people', he stated. 'That social norm is just something that has evolved over time.'[9] Zuckerberg argued that the rise of social media reflected changing attitudes among ordinary people, adding that this radical change had happened in just the six years since Facebook was founded. The continued use of social media over those six years had fostered a new culture of openness and connection, according to Zuckerberg, attributes to be valued above privacy and secrecy. On another occasion, he proposed that before the advent of the internet, we suffered 'privacy through obscurity'. The development of the internet, Google, Facebook and other social media has constituted a 'democratising' new form of expression, lifting us out of obscurity and allowing any individual to publish and share in public.[10]

Yet Zuckerberg's stated preference for radical transparency over obscure privacy has been blocked, or at least checked, on the occasions when Facebook has attempted to change its privacy protocols – changes resisted by Facebook user groups. The revenue of Facebook, as of Google, derives largely from advertising. In effect, the Facebook users are the audience sold to advertisers; the information available on users' 'likes', preferences, habits and desires is invaluable to advertisers – and the greater access they have to that information, the more valuable it becomes. In this regard, 'Facebook still sees privacy as an encumbrance that could compromise potential profits', according to Simon Davies, director of Privacy International in the UK.[11] The openness of social media users entails an openness of their personal information to advertisers, who can then target users individually as well as collectively. As was observed of Google in

2011, the many millions of Android mobile phone users have their every movement tracked by Google, which makes more than 95% of its revenue from advertising: all that stored information on users' movements and lifestyles is the 'gold' that Google offers advertisers.

Zuckerberg's contention that attitudes to privacy have shifted since 2004 and the advent of social media is supported by other observers. The journalist Emily Nussbaum wrote in 2007, in a study of the internet 'and the end of privacy', that privacy was the grounds of a new generation gap. The parents' generation is appalled by their teenagers' disregard for privacy; those teenagers are seen as having 'no sense of privacy' and a dangerous commitment to posting all aspects of their personal lives, including erotic photos, online.[12] Sherry Turkle's detailed study in 2011 of social media users confronted the issue of privacy. Turkle concluded that while teenagers and young adults are portrayed by the media as no longer caring about privacy, her finding was different, if equally disquieting. Turkle gleaned from interviewing young social media users that they genuinely did not understand the rules of privacy. They didn't understand privacy protocols on social media sites; they were unsure if surveillance was legal; they didn't know if they were entitled to any protections. As a result of this general ignorance and confusion, many users had a resigned acceptance to the likelihood of surveillance and other intrusions on their privacy.[13]

Privacy, then, has become a troubled space. Whether the subject of ignorance or the object of disdain, it is no longer universally held as an inviolable social right. It has been construed as an unimportant form of space, even a malignant space – in the view of at least one journalist. Privacy was once upheld as a core value of citizenship and of fair and just civil societies. The space accorded privacy, however, has shrunk noticeably in the age of the internet. How has privacy come to this?

A brief history of privacy

Privacy is a recent invention: there is little trace of it – or concern for it – before the eighteenth century. Throughout human history, most cultures have had a very limited notion of privacy; many had no words for 'privacy' or 'self'. For hunter-gatherer societies, the community prevailed over the individual; there was no space, and no desire, for solitude. To be alone was reserved as a form of punishment: social exile, removal from the group. This alone-ness equated to social death, or simply death. Exceptions to the communal rule occurred when a young individual was required to spend time on his own as an initiating rite – but this separation from the group was only ever a temporary solitude, before the group welcomed back its newly initiated member.

The five-volume study *A History of Private Life*[14] details the minimal space of privacy in the ancient world. The growth of towns and cities did not bring an accompanying space for individual solitude. The family household was crowded with dependant relatives and slaves, as well as family members. Perhaps the sole exception to the emphatically communal social norm was the intense spiritual solitary experience – in the desert or wilderness – undertaken by holy men, shamans, prophets or visionaries, who then returned, inspired, to benefit their communities.

In the European middle ages, people lived together, slept together, ate together, travelled together: life was always 'surrounded', involved with others. Residences contained no room for solitude; the family was generally extended, providing more agricultural labour or domestic help. The household was crowded, and privacy did not exist. Even literacy served a communal purpose: readers in towns read out loud from manuscripts to audiences. The lowly status of privacy in societies which valued community is preserved in the roots of the word itself: the Latin *privatus* meant 'withdrawn from public life'; the word derived from *privare* meaning 'bereave, deprive'. To be disconnected from public life was a state of deprivation. The stigma formerly associated with privacy is reflected in the military rank of private: the lowest rank.

The invention of privacy occurred as part of a huge economic and social shift, most evident in eighteenth-century Europe. As charted by Jurgen Habermas in *The Structural Transformation of the Public Sphere*, the rise of the bourgeois class introduced a new distinction between public space and the private home. A social emphasis on the private sphere, initially centred on family life in the home, was first noted in the Netherlands, a century before a similar shift occurred elsewhere in Europe. Holland in the seventeenth century was Protestant, upholding values of individualism, order and thrift. The Dutch republic had no landless peasantry, and possessed a strong mercantile class; most of the population lived in towns dominated by the new middle class: this was the first bourgeois state.

The Dutch had a limited use of servants, expecting the middle-class housewife to run the household. The newly established domestic familial space was clean, ordered, quiet, feminine – as reflected in the paintings of Vermeer. These famous works portray a new interior mode: a still, quiet environment, 'a setting for private acts and personal moments',[15] especially as performed by women. This defining of the domestic space as the site of privacy, to be valued and guarded, was repeated across Europe in the eighteenth century, as the middle class asserted its social power. Privacy and the importance of the home became core values of the Bourgeois Age, which was founded on 'domesticity, privacy, comfort, the concept of the home and the family'.[16] The historian Phillippe Ariès, in his social history of family life, writes that the middle class in Europe of this period 'began to hold society at a distance, to push it back beyond a steadily extending zone of private life'.[17] The public sphere, conducted in coffee houses and other social spaces, was a feature of this age; but the invention of the domestic sphere as a zone of private space was equally significant.

In the domestic sphere, private residences were built according to new principles. Privacy was promoted in the building of corridors, passages between one private room and another. Whereas the medieval household was crowded and had beds strewn throughout, the newly constructed bourgeois home had private bedrooms, spaces designed to enhance the privacy and domestic intimacy favoured by the bourgeois family. As the family itself continued to shrink – extended family members and grandparents removed from the household in favour of the nuclear family – the core family members enjoyed still more personal space.

Ultimately, privacy came to be centred not on the family but on the individual: sheltered in his or her own bedroom, free of unwanted intrusions onto personal space. There were other cultural factors fostering an emphasis on the individual self. Protestantism rejected ritual, even the dialogue of confession, leaving the individual alone with their internalised God. Reading too became silent and interior, creating an inner space that is the reader's alone. Industrial capitalism focused on the individual as the unit of production and consumption, further weakening the family as the core social unit. Privacy became enshrined in law as an individual right – the right to be left alone, in the nineteenth-century definition. Private space – once unknown and unwanted – had come to be cherished as a protective ring around the inner core of self.

True confessions

If the private and the public began to separate in the eighteenth century, how were the barriers between them erected and supported? How was it determined whether an act, expression or behaviour was fit for the private sphere rather than the public? A gender division operated within societies organised along patriarchal lines: the public sphere was the domain of men, discoursing in the coffee houses and other public sites; the domestic sphere was for women, at home and in private. A corresponding distinction emerged between the serious, political content of public discourse and the emotional, personal matters expressed in private. The public sphere was the site of weighty matters of political importance, while the private was the realm of intimate, emotional consideration – conducted within the walls of privacy and expected to stay there. A diary, for example, was the personal record of one's emotional life, intimate thoughts that were not fit for publication.

Gradually, this division between the social public and the intimate private began to break down, so that intimate thoughts were eventually expressed in public. A major turning point in this direction was the publication of a book at the end of the eighteenth century by Jean-Jacques Rousseau, the most famous French writer of this time. Rousseau's *Confessions*, published posthumously in two parts in 1782 and 1787, was a landmark publication in many respects: the first truly autobiographical literary work, the first in the 'confessional' mode of literature, the first book to concern itself not with the author's actions or thoughts, but primarily with the author's feelings. Rousseau's book exposes, with an openness never before seen, his feelings of guilt and shame resulting from his theft of a ribbon at age 16, a crime which he blamed on a household servant. Tormented by a lifetime of guilt that he may have ruined the life of an innocent girl, he nevertheless had never confessed his crime, even to his most 'intimate friend'. His public admission of guilt, in *Confessions*, was a major factor in his writing of the book, as an attempt to relieve the burden of his shame.

Another feature of Rousseau's *Confessions*, shocking to his peers, was his revelation of the quirks of his sex life. Rousseau detailed his fantasies and 'often-embarrassing' behaviour, including masochism, exhibitionism, masturbation and ménages-à-trois. So frank were his revelations that Rousseau himself found these confessions the hardest to make. Critics attacked *Confessions* not only for its lurid content but also for its very nature as a candid autobiography. 'We are often disgusted at that self-importance of this honest man', wrote one reviewer, 'which makes him think that every little story that regards himself... must be interesting to the public'.[18] Few bloggers and memoirists today worry about such lack of interest – but the reading public in 1782 was unfamiliar with the confessional mode.

The Romantic movement took many of its cues from Rousseau, including his elevation of feeling and emotion, and his emphasis on authentic expression of the self. The nineteenth-century Romantics added the cult of genius, in which the highest artists were thought to project qualities of the inner self – sublime, often tormented, touched by madness – into their art in a heroic struggle of public expression.

Rousseau launched the confessional mode of public intimacy, in which an individual makes public penance of a private sin – which becomes public knowledge through publication. This mode of writing – expressed through a wide range of media platforms – became prevalent in the late twentieth century, but the nineteenth century was less fertile ground for emotional journeys lived out in public.

A check on the growth of public intimacy occurred in the later nineteenth century: the Victorian period restrained public displays of emotion by enforcing strict moral codes on social behaviour. The rise to supremacy of the middle class ensured that bourgeois values of thrift, hard work and moral decency became the ascendant cultural values. The private sphere was consolidated as the domestic space of family life, while the emotional and personal were largely repressed in public, confined to secretive private currency. The dominant media form for the Victorians was the newspaper, the dominant entertainment form the novel. The inner world of reading was often enchanted by novels probing characters' inner secrets. Plots of vast Victorian novels turn on the exposure of characters' secrets – shameful secrets that they have managed to keep for a lifetime but which, when exposed by a villain, could spell social doom.

From rear window ethics to Facebook: Privacy and technology

Public intimacy escalated in the twentieth century, facilitated by developments in camera and broadcast technology. *Candid Camera* began on television in the US in 1948, featuring everyday citizens filmed in embarrassing acts for the enjoyment of viewers. The telephoto lens was a significant addition, allowing spying – and photographing – from a distance. Alfred Hitchcock's great film of 1954, *Rear Window*, revolves around the properties of the telephoto lens and its breaching of the private intimate sphere.

The central character Jeff, played by James Stewart, is a professional photographer, temporarily immobilised in his apartment, his adventure restricted to the power of the look. He is a man without legs, but with a techno-extension of his eye. When the villain comes for him at the end, his only protection is flashbulbs. He is man as camera. The windows of neighbouring apartments that he peers into are like movie screens: some holding entertainment for Jeff, but one disclosing the narrative of a possible murder. Jeff is a voyeur, fascinated by the window narratives before him – but so is Hitchcock, and so are we.

At one point in the film, Jeff agonises out loud over the rights and wrongs of spying on a neighbour. 'I wonder if it's ethical,' he ponders, 'to watch a man with binoculars and a long focus lens.' His girlfriend Lisa (Grace Kelly), scornful of his viewing habits at first, can only answer: 'I'm not much on rear window ethics.' These rear window ethics – the morality of looking – is the film's central issue. *Rear Window* was one of the first major works to plumb this modern ethical fissure – the one opened up by the meeting of camera technology with the innate curiosity about the lives of others.

The film dramatises the moral ambivalence of looking at others through cameras: most of the characters in *Rear Window* find the practice objectionable, but are drawn to it nevertheless. Lisa comes to share in Jeff's voyeurism, which had first repulsed her as a 'diseased' obsession. She rebukes herself as well as him when she says, 'We're two of the most frightening ghouls I've ever seen.' Jeff's friends declare his obsession with snooping perverse. A detective friend warns him: 'That's a secret private world you're looking into out there.' The nurse Stella is equally disapproving. At first she is staunch in her old-fashioned morality, which regards voyeurism as a sin. 'The New York State sentence for a peeping tom is six months in the workhouse', she tells Jeff, then adds: 'They've got no windows in the workhouse.' She can sense trouble coming as a result of Jeff's sin. 'You know in the old days they used to put your eyes out with a red-hot poker', she warns him. Yet before long she herself is seduced by the rear window spectacle. She is no better than anyone else at resisting temptation. 'Oh dear', she laments, 'we've become a race of peeping toms'.

Hitchcock's moral world-view, which could best be described as English-Catholic-Freudian, expressed itself with suitable ambivalence in *Rear Window*. Hitchcock himself was a master peeping-tom, and his films are licensed voyeurism. His hero in this film is a man with a camera: he is the enemy of privacy. Jeff is driven by the lure and the promise of the look. It gets him into trouble, it almost gets him killed, but he can't resist its force. Rear window ethics don't prevent him from spying, first with binoculars then camera lens, on his neighbours. Jeff is justified in the end for his voyeurism, because it catches the murderer. Yet his unbridled acts of prying cannot go unpunished according to the morality of the film: he ends up with a second broken leg, leaving him even more immobile. He does, however, also end up with Grace Kelly.

The mid-twentieth-century rear window ethics on display in this film are prompted by the properties of the camera. The technology of the camera, the key component of film and television, and magazines, allowed viewers to see others more closely: first those in public life, then the private lives of whomever the camera was aimed at. The distinction between public and private became blurred: anything shown on a screen, or in a magazine photo, no matter how private, was public property. The object of the camera was there for the communal gaze; the camera beckoned us to look, and we looked, eagerly.

As Joshua Meyrowitz detailed in his study of television, *No Sense of Place* (1985), the camera is a technology that exposes. It creates a culture dependent on exposure: the private life of presidents and other celebrities, the nervous tic of a guilty party in television close-up, the domestic life of a star – all brought to us by telephoto lens. The telephoto lens allowed widespread prying into the private lives of celebrities, in the hands of paparazzi – a term, meaning annoying buzzing insect, coined in Fellini's *La Dolce Vita* of 1960. Paparazzi photos became the staple of magazines selling images of celebrity – usually against the celebrities' wishes. Much of this magazine culture is exposure for its own sake, the downfall of privacy. The public are willing voyeurs; the objects of their gaze may be willing or they may be not: it makes no difference. Nurse Stella in *Rear Window* was right when she says we've become a race of peeping toms: we are all rear window snoops, via the spying powers of media technology. The telephoto lens attached to a camera publicised worlds that were previously hidden behind walls of privacy. The camera could now see through or over those walls, rendering the previously private into public property: images in magazines or on screens.

As popular culture became increasingly visual, and increasingly saturated by cameras, some celebrities sought to resist the intrusion and maintain a private space, while others did the opposite, offering up views of their intimate self and relationships in books and films. Madonna invited us into her bed in 1991 in the film *Truth or Dare*, allowing us to see not only her domestic sphere, but also to hear her private complaints and anxieties concerning her fabulous lifestyle. One of the truths disclosed in the documentary was that the celebrity at its centre 'doesn't want to live off-camera'.

The advent of the World Wide Web in 1993 offered a route to democratise public intimacy. Now you didn't need to be a celebrity to let everyone know your intimate feelings and relations. Jennicam, operated by Jennifer Ringley from 1996 to 2003, was a pioneer in the show-all outpouring made possible by the internet. Jennicam's motto was 'I like being watched', and Ringley's deployment of the webcam exposed her intimate life in great detail, including her sex life and relationships, but also her concerns and emotional life.

Jennicam was a precursor to reality TV of the *Big Brother* type. *Big Brother* first aired in 1999 in the Netherlands, before taking over the commercial television world. In the twenty-first century, nations – Western and non-Western – and particularly their youthful component, appeared to celebrate the visual, the personal, the demotic, the intimate, the banal, even the boring, aspects of the everyday life of everyday citizens, live on television for hours on end. Culture was about surveillance but also performance. Surveillance technology, once central to the dark and harrowing vision of a totalitarian society, as found in George Orwell's *1984*, was now central to the quest for momentary celebrity by willing participants on *Big Brother* or the many other reality TV shows.

The contrast between the two versions of Big Brother is telling: Orwell's novel, published in 1949, warned of the loss of privacy in a totalitarian state deploying surveillance technology to dominate its citizens. 'Big Brother' is the personalisation of this sinister state and its oppressive apparatus of vision; a Stalin-like paternal figure, Big Brother is a projection of parental benevolence masking a malevolent authoritarian state. *Big Brother* the TV show, by contrast, is breezy entertainment in which young volunteers willingly expose themselves before all-seeing cameras within a specially constructed house.

The Orwellian world in *1984* is gloomy, dystopian. The reality TV *Big Brother*, by contrast, is bright, sunny, youthful. The Orwellian view presents a bleak vision of corroded humanity; television reality is cheery, chatty, hedonistic. For Orwell, surveillance technology is overt, oppressive, sinister; for *Big Brother* contestants, the surveillance technology is covert, but assumed and readily accepted. In the Orwellian world, cameras and surveillance ensure a dread, closed society. The contestants on *Big Brother* instead embrace surveillance technology as a gateway to television exposure and temporary fame as a television celebrity.

This contrast indicates a broad cultural shift concerning the status of privacy. Previously, privacy was considered a precious right of individuals, to be prized, protected, and guarded against erosion by government or corporation. Now, privacy is increasingly given up or sold – willingly – by individuals, in the service of attaining media exposure. That is, privacy has become a commodity, traded by its owners in return for a brief spot-lit moment in the 'reality' of fame-generating television. If contestants have a secret, it must be exposed, on television, for all the viewers, and then on the internet, for many more viewers. That is the contract of reality TV: participants give up their secrets, their intimate moments, their private relationships, for the camera – which makes the previously private into public property.

Television is licensed voyeurism of real life, as displayed in its staples: news, current affairs, reality TV. The continuous prying into grief, violence and personal lives in these programmes provokes a few objections about the invasion of privacy – but the ratings are huge. Reality TV shows attract a never-ending supply of people willing to expose themselves before the camera. Whether for exhibitionism, money or the thrill of being on television, countless individuals gladly give up portions of their own privacy. Reality TV wouldn't work unless people were willing to surrender the very privacy once held sacred in Western societies.

Reality TV entertainment, based on the ever-present surveillance camera, became a staple of television programming around the world at the same time as closed circuit television (CCTV) cameras proliferated in cities. Used by governments, councils and private property owners to surveil property and public space, CCTV has become ubiquitous in most urban spaces. The British theatre director and theorist John McGrath, in his 2004 book *Loving Big Brother*, proposed that the prevalence of these cameras – and the knowledge of their prevalence – has created a 'surveillance space', in which individuals perform for cameras in public spaces, just as reality TV participants perform themselves before the cameras.

The final element in the recent cultural shift to public intimacy is social media, or Web 2.0, dating from 2004. Webcams, blogs and social networking have added another dimension to this dynamic: the compulsion to connect with others. Radical transparency online includes public intimacy, as users share private thoughts, feelings and experiences, and intimate photos, with each other on the network. The confessional mode of expression inaugurated by Rousseau has become a staple of the extended media sphere of the twenty-first century. Public intimacy is everywhere, across multiple platforms: blogs, websites, published autobiographies, radio and TV shows, Facebook pages, Instagram and countless social media sites.

American television talk shows came to love the confessional in the later twentieth century, whereby a celebrity found guilty of misdemeanour must immediately confess and repent – in public, on television – and hope for public forgiveness. The redemption narrative inflects the confessional as a form of public intimacy. No matter how great the crime, there is a chance that the sins may be forgiven, so long as confession is made in public, before the cameras, and with sincere-seeming tears. Social media has made this possible for anyone, as 'the people formerly known as the audience'[19] produce online content as well as consume it. Much of this content is of the intimate, emotional sort that was once deemed private, not for public exposure. Today, the intimate is regularly made public.

And what of privacy itself? Civil libertarians, civil rights groups, privacy advocates and public custodians such as privacy commissioners worry repeatedly – and publicly – that privacy is dissolving, melting away in the age of ubiquitous cameras and information. Certainly the space occupied by privacy is now a diminished space, severely reduced from its grand dimensions enjoyed in the nineteenth century. Other social priorities have emerged which have shrunk privacy's status – such as the need for security, as constituted by ubiquitous CCTV cameras, and the desire to connect on social media – overcoming much of the reverence once relished by privacy.

A news item reported in 2012 encapsulates these priorities, as well as the devaluing of privacy by many citizens. In Sydney, a dog named 'Bucky' was stolen, but footage of the dog-napping – showing the culprit – was captured on CCTV.[20] The owner of the dog obtained this footage and posted it on Facebook; soon the teenage culprit was found and 'Bucky' returned to its owner. The owner told the media that in this instance CCTV and social media had combined to solve the crime, proving that 'someone's always watching . . . you can't get away with anything'. The Deputy Privacy Commissioner of NSW did not support this moral, however, telling the same newspaper: 'It's concerning that people would post images of children online without the consent of a responsible adult . . . members of the public shouldn't be taking matters into their own hands.' The owner of 'Bucky' disagreed: 'I think we're a bit beyond privacy in this day and age aren't we? No one's got any privacy.'

7

Photography 2.0: Photos on the Loose

The distribution of digital information across the internet has taken many forms, and has effected some major cultural shifts. The downloading of MP3 music files through peer-to-peer sites has offered free access to music, decentralised the music distribution business, defied copyright regulation and transformed the music industry and – to a large extent – the social role of the musician. Yet music files are only one form of digital information coursing across the internet. Digital photographs are another; the status, impact and social significance of photographs have also altered with the advent of networking.

The cured image

In his 2010 essay 'Two Cures', art theorist and historian Boris Groys investigates the complex relations between traditional analogue and contemporary digital images – and also the relationship between images and curators. Groys elaborates on the theme of the curator as someone who 'cures' images, allowing these otherwise 'helpless' entities to be represented in exhibitions. The work of the curator permits spectators to be led to the artwork, 'as hospital workers might take a visitor to see a bedridden patient'.[1] He contends that the digitising of images is also their liberation from institutionalised frameworks, constituting a new form of cure.

This recent liberation of images, from the hospital walls of galleries and museums, has ensued in the wake of digital technology and, more importantly, global networks of circulation. Groys claims that digitisation and networking have allowed 'images to become independent of a curatorial practice of any kind, to become really strong'.[2] Images are no longer sickly and helpless, needing to be assisted onto white walls by (admittedly well-meaning) curators. Digital images are now genuinely robust, strong and healthy. They do not need the closed space of an exhibition in order to be seen and appreciated. As Groys observes, they 'have an ability to originate, to multiply and to distribute themselves through the open fields' of networked communication. There is no need for 'centralised curatorial control'; digital images flourish without support, depending only on 'their own vitality and strength'.[3]

The new environment in which this liberation of images has taken place has been formed by the convergence of communication technologies evident in the realm of Web 2.0. Certainly the old mass communication model of broadcasting from central institutional or corporate points of transmission to passive consumers has been left behind. The term 'prosumer' has been activated to embody this new world of communication, in which an individual may now be a producer as well as consumer of information. Any individual may now self-publish by means of a blog, Facebook or Instagram page, or Twitter account. Convergence means that almost everyone carries around with them a device that is simultaneously a phone, a computer, a camera, a GPS navigator and a means to access the internet. Any individual carrying one of these devices has the means to be a citizen journalist.

With reference to photography, any individual with a camera phone or digital camera has the means of being a photo-journalist; any individual can publish their photographs in an online blog or virtual gallery. Instagram and Pinterest have been used to mount virtual galleries of digital photographs. The American artist Richard Prince has celebrated picture-sharing sites as mobile miniature museums: 'It's like carrying around a gallery in your pocket.'[4] Much of this convergence and transmission/reception of information occurs wirelessly, and increasingly large bodies of digital information are archived 'in the cloud', that is, in networked repositories of computer servers, with no need for hard or terrestrial copies. The cloud – also known as the celestial jukebox, celestial cinema, celestial gallery, celestial library – stores enormous reservoirs of music, movies, still images or text, to be accessed by users and subscribers from any point, including from portable devices.

The gatekeepers attending the traditional means of publishing – editors, curators, publishers – are absent in the cloud, allowing greater likelihood of public exposure of an individual's photographic images. The builders of the digital networked future – including Google – are seeking to enhance this gatekeeper-free access to cultural expression. Google's Cultural Institute in Paris is a laboratory of the digital image, with 650 cultural partners around the world. Its director stated in 2015 that the aim of 'Le Lab' is to predict – and make possible – 'the collective experiences with digital content in a physical space'.[5] Google's online galleries of visual artworks can be accessed by users to curate their own virtual galleries. Over 350,000 online galleries have been created; in this manner Google is facilitating a further 'cure' of the image, with the advent of the citizen curator of images.

The political image

If it can be claimed that the digital image has been liberated from the clutches of curators, this same process has occurred, in much more serious contexts and with profoundly serious consequences, in those regimes that have exercised control of information and images through censorship and state regulation. The Arab Spring began in Tunisia in December 2010, and social media – Facebook, blogs, Twitter – featured strongly in the organisation of protestors and the promulgation of their cause.

Photography played a crucial role in promoting resistance to an oppressive government, in documenting civil protest and the violent responses by police and security forces. Photographs distributed on social media performed the dual role of galvanising the resistance movement through documenting the outrages of a violent state response and – because the internet is a global network – publicising the political upheaval around the region and around the world. After the flight from power of the Tunisian president Ben Ali in January 2011, despotic leaders were removed in Egypt, Libya and Yemen, while uprisings and major protests occurred in Algeria, Iraq, Jordan, Kuwait, Morocco and Oman.

The same boosters and enthusiasts who cheer online journalism and all things shiny, new and networked have also celebrated the Arab Spring as the direct consequence of Web 2.0: the various uprisings have been dubbed 'the Facebook revolution' and 'the internet revolution'. This simple Silicon Valley determinism needs to be resisted if it fails to register the pressing social context which lay at the base of these uprisings and protests. Tunisians, for example, had experienced oppressive living conditions for many years, including high unemployment, government corruption, rampant food inflation and restrictions on freedom of speech. The first Arab Spring revolt did not occur out of the blue, not even the Facebook blue. Dissident activists and union organisations had worked as a resistance movement for many years, and there had been civil protests before – most recently in 2008, when rallies and strikes, organised in the mining area of Tunisia, were suppressed by government forces.

The difference with the events of late 2010 was the scale and speed of the protest movement, its ability to unite disparate groups against the government and the widespread dissemination of its actions. In this regard, photography was prominent. As is well documented, the catalyst for the Arab Spring was the self-immolation of Tunisian fruit-seller Mohamed Bouazizi on December 17, 2010. This protest against corruption and oppressive living conditions became the new rallying cry for a range of groups, including political dissidents, human rights activists, trade unionists, students, lawyers, academics and the long-term unemployed. People from Bouazizi's home town of Sidi Bouzid created a Facebook page called 'Tunisian Citizens Are Setting Fire to Themselves Mr President'. Social media became the means of transmitting dissident information, as the mass media was largely in control of the government and failed to report anti-government demonstrations and protests.

One of the Tunisian internet activists was Lina Ben Mhenni, who shared dissident news on her blog 'A Tunisian Girl and Her Facebook Page'. She went further than simply relaying information: she attended the demonstrations, first in Sidi Bouzid and then in other towns; she took photos and videos of the events, which she posted on her blog. She visited hospitals, photographing victims of violent responses by police and security forces. These photographs, regularly added to her blog, documented the rapidly escalating protest movement, while – more dangerously for herself – publicising to the world the brutality of the Ben Ali regime against its own people.

Lina Ben Mhenni later recounted her activities as 'A Tunisian Girl' at a 'Networked Revolts' public event in Greece. She noted that the first demonstration in the capital Tunis was small and ignored by the traditional media. Bloggers and cyber-activists constituted the only documenters of the protest. At a second demonstration in Tunis two days later, the police reacted with violence, beating and injuring protesters. She continued to cover the demonstrations in Tunis, especially the strikes by lawyers, who were active in these events. One of the lawyers was kidnapped and beaten by the police. Ben Mhenni stated that despite the violent suppression of the protests, demonstrators were not deterred, and the demonstrations expanded in size.

A protest event on January 8 drew a large group of demonstrators; similar events occurred throughout Tunisia. This escalation of the movement 'infuriated the police who became even more violent';[6] demonstrators were killed in various cities throughout Tunisia. On January 9, Ben Mhenni left Tunis to cover a demonstration where the police reacted with extreme violence. In Regueb, five peaceful demonstrators were shot dead by police. She took photos of the deceased and published them on the internet, receiving hundreds of comments and contact by foreign media wishing to cover the unfolding events in Tunisia.

The scale of the uprising was now becoming irresistible. On January 14, Tunisians across the country marched in the streets shouting 'Ben Ali out'. Police responded with teargas and more live ammunition, killing more protestors. But the president fled the uprising, finding exile in Saudi Arabia and ending 23 years in power. Police violence did not relent, however, even after the president's flight. Protests continued against the newly declared state of emergency and the caretaker government, which included members of Ben Ali's party.

It is not impossible for an authoritarian government to suppress social media and the internet. Bloggers in Tunisia were kidnapped and beaten for their postings, and Ben Mhenni's Facebook account was censored and her university wage frozen. She has also received death threats. But the dissident information, and the accusatory photographs, emerged nevertheless on the internet. While the press, radio and television transmitted government propaganda, the social media leaked an alternative view. Ben Mhenni herself is ambivalent on the role of internet activities within the protest movement. She rejects the description 'Facebook revolution': 'it started on the ground', she insisted in a 2012 interview, 'when Mohamed Bouazizi set fire to his body, and the people started to demonstrate and were attacked by the police'. It cannot be called an internet revolution, she has argued: hundreds of people lost their lives, and many more were injured. 'The internet just played the role of the mass media when the journalists couldn't do their work here.'[7]

Yet at the 'Networked Revolts' event, Ben Mhenni acknowledged the significant role played by social media and the internet in Tunisia. She pointed out that 'against the blackout attempted by the traditional media', not only did social media 'cover the uprising, but they also helped in mobilising people to participate in the demonstrations'. She contrasted the successful uprising with that of 2008, when the regime succeeded in suppressing social unrest. One of the reasons was that in 2008 'only a slim minority were using Facebook, blogs and social media ... the Tunisian government simply ordered the police to arrest the leaders of the movement, they put them in jail and the movement disappeared'.[8] Three years later, the same regime was unable to contain the spread of information – and photographs – across social media.

Certainly the euphoria attending the Arab Spring across the region had dissipated by the middle of 2012. A period of instability ensued, characterised by violent responses to protest movements, military takeovers of government, repression by authoritarian governments, internecine religious conflict and pro-government militia violence. So severe was the aftermath to the Arab Spring in many nations that some commentators have dubbed the succeeding period 'the Arab Winter'. Yet the protest movements continue to operate, in Tunisia and other nations, traversing the internet through social media. Photography continues to document, represent and galvanise protest. These photographs are published not only in newspapers and magazines, but – often with far greater reach and far greater impact – across the internet.

Fried eggs and social media

Online communication in the wealthy Western nations does not often engage with the heightened political consciousness seen online in Tunisia and elsewhere. Indeed, much of social media communication – including the transmission of digital photographs – dallies with exhibitionism, narcissism, solipsism; all the less desirable -isms populating the World Wide Web. The obsession with 'selfies', crowding Instagram and other social media sites, has been derided as a contemporary narcissism; the 'selfie sticks' used to take photos of the self in public have been termed 'wands of narcissism' and have been banned from some rock music venues in Australia.[9]

Yet social media also involves sharing, collaboration, collective authorship and the navigation of new constructs in time and space opened up by convergence and networking. Photography is deployed within these networks in new modes of expression, communication and even identity formation.

In her book on the uses of social media, *Alone Together* (2011), Sherry Turkle notes that blogs and Facebook have created new online spaces filled with interaction and archiving – of personal and even family histories. This is where photography is frequently at play. Photographs have always been integral to Facebook, since its unruly origins as 'Facemash' – for which Harvard student records were hacked to create files of paired photos of students, inviting users to nominate which one was 'hotter'. This undergraduate enterprise established the brazen disregard for privacy and accepted codes of propriety, and political correctness, that later made Facebook the greatest success story of Web 2.0. It also established the practice of representing – in this case objectifying – individuals by their online photo profiles.

Turkle has investigated the means by which profile photos – and other images displayed in photo galleries – have contributed to the creation of the online self, which she has elsewhere characterised as 'the second self'. The selection of photos displayed on a Facebook page constructs a version of the self, as one of her interview subjects expressed it: 'you're making it up for all these people… You can create who you want to be'.[10] Facebook or Instagram albums constitute an edited life, projecting an ideal self to the world via the network.

This construction of an idealised self can involve Photoshop, or the manipulation and 'improvement' of one's image. This process is facilitated by apps and services including PhotoWonder, which has 'slimify', 'blemish fix' and 'make-up' options; Facetune to erase wrinkles; and PicMonkey to whiten teeth and apply a 'spray tan'.[11] The effect of this digital work is the building of a 'second you', a 'twin on the Internet', as one of Turkle's interviewees describes it.[12] The main profile photo is an avatar of the self, often chosen with great deliberation, while other photos, selected, arranged and possibly digitally altered, project aspects of the self in the best possible light.

Turkle notes one other facet of this representation of the self through photographs, or photographic performance of the self. That is the ongoing, temporal nature of this performative act. Different versions of the self – performed through the selection of different avatars and different combinations of photos – may be tried out over time. In this process, the most important factor is the response of friends. A negative reaction from the network of friends will produce an edit of the selected photos, a re-arrangement of the constellation of self. Here the social interaction aspect of social media is pre-eminent. Social media, including Facebook, is built on sharing, relaying information, connecting with others. The performance of the self on Facebook functions within this apparatus of sharing. Turkle calls the self enacted through social media 'the collaborative self',[13] composed of responses to reactions from others on the network.

A similar observation is made, from an anthropological perspective, by Deirdre McKay in her 2010 article 'On the Face of Facebook'. Primarily concerned with the use of photos on Facebook by a network of young Filipinos, McKay was struck by the frequent use of historical images by this group of friends. These imported images were of ancestors and past events, either of personal or wider, collective significance. The representation of the past in this way was for McKay a distinctive use of the 'compulsion to visibility'[14] urged by Facebook and social media, one not found to the same extent in Western cultures. Yet McKay and other contemporary anthropologists have claimed that social media is mapping, and perhaps enabling, a new conception of the self. The use of photos by the Filipino friends reveals 'extended and relational personhood'.[15]

But all users of Facebook photos, anywhere on the global network, partake of this same practice of the collaborative self or relational personhood. Western societies have for centuries emphasised and rewarded individualism, whereas Facebook fosters a sense not of the isolated individual but of the dividual – the divided or connected self. As one anthropologist has noted, 'Facebook offers not a new form of personhood in itself, but reveals potential for something akin to Melanesian dividuality.'[16] Even if Facebook only highlights a realisation of the fundamentals of personhood – through connection to others – this 'suggests a profound change in Western societies'.[17] McKay relates a metaphor suggesting the difference between Western and non-Western conceptualisations of the self: Westerners are 'individual fried eggs whose edges do not touch; Filipinos are eggs fried together so that their whites blend, leaving a pattern of yolks embedded in a wider field.'[18] The emphasis on connection and collaboration fostered by social media may be altering the Western egg.

This process perhaps can only go so far. The Facebook profile photo is primarily an avatar of the individual self. Users may select a flattering version of themselves, or an enigmatic one, or an ironic disavowal of the representation of self through an amusing substitution of object for personal image. Selfies record, as irate critics have claimed, 'the ultimate narcissism'. But no matter what photos are chosen for display on Instagram, Facebook or other social media site, those images are selected for display to a network of friends and others, suggesting the importance of collaboration and interaction. The Western egg may not ever be fried together, or scrambled, but it may begin to run a little.

8
Schizochronia: Time in Digital Sound

You are sitting in front of a screen. The screen may be part of a digital audio work station, or it may be the screen of a laptop. You press a key and watch as the cursor moves through the waveform. You hear the sound at the same time as you see it traversed by the cursor. You decide to retrieve a sample, which you've stored in the computer. It's located way up ahead of the present waveform: in a few seconds you've scrolled forward, claimed the sample and positioned it next to the waveform. You magnify the image, to get a better 'look' at the sound. You decide to insert the sample into the waveform, trying various positions. If you change your mind, nothing is lost: this is, after all, non-destructive editing.

This chapter examines digital audio technology – including sampling, editing and mixing – as an 'intellectual technology'. In Pierre Levy's words, such a technology modifies the 'cognitive ecology' into which it is introduced.[1] Digital audio shares the properties of other digital technologies: it is founded on the immaterial (information); this information is endlessly manipulable. These properties form the basis of an intellectual economy, in which familiar epistemological categories – including intellectual property and the codes of realism – are challenged. Digital audio adds to these general properties its distinctive re-alignment of the technological reproduction of sound. The digital audio user works with sound in a specific manner, particularly with regard to time. The focus of this chapter is on the consequent refiguring of our conceptualisation of sound, in terms of time.

Mathematical time

Digital audio is fundamentally a numerical technology. Unlike analogue audio, which creates an analogue of the waveform in various media (voltage control, deviations in a groove, magnetic patterns on tape), digital audio represents a sound event as a set of numerical values. This binary data is processed and stored – as information – to be reconverted to the original waveform at the point of output. The most cited advantages of the digital process are its lack of degradation in copying, its ability to error-correct and its flexibility in editing. All these advantages derive from the fact that digital audio works with an immaterial stream of data – with numbers.

Discrete time sampling has been called 'the essence of digital audio'.[2] This technique encodes the analogue waveform into infinitesimal pieces of information. Each slice is discrete in time; the standard 44.1 kHz sampling rate means that 44,100 time-samples per second are taken of the waveform. An analogy used to describe this process is the technique of cinema: the 24 discrete frames per second of film merge to reproduce images of movement. Digital audio takes 44,100 discrete snaps per second, to reproduce variations of sound in time.

Such a description of this technology invokes Bergson's philosophical treatment of time (discussed in more detail in Chapter 2). Writing at the beginning of the twentieth century, Bergson drew a distinction between time as constructed by the intellect, and duration as glimpsed by intuition. The former, which he variously termed scientific or mathematical time, is a succession of instants divided by the pragmatic faculty of intellect. In a famous analogy, he likened the workings of intellect – cutting up movement into discrete moments – to the cinema apparatus, taking 'snapshots ... of the passing reality'.[3] Each frame represents a static point, isolated from the flow of movement: this is the 'cinematographic illusion' perpetrated by intellect. It is only through intuition that this illusion can be overcome, that 'the infinite multiplicity of becomings'[4] can be grasped.

Yet in his *Cinema 1: The Movement-Image*, Deleuze salvages the cinema from Bergson's 'rather overhasty critique'[5] appraising the artform in Bergsonian terms. At the same time, he deploys cinema as a mode of thought, with its own concepts. In the same way, digital audio may be approached, from a Bergsonian perspective, in more than one way. The very basis of digital audio – its 'essence' – is founded on the binary number system developed by Leibniz – whose thinking is a cornerstone of the 'universal mechanism' assailed by Bergson. Mechanistic science, as characterised by Bergson, took 'time as an independent variable', to be measured, to be divided into ever smaller intervals: 'as great a number of moments as we wish in the interval of time it considers'. As a result, according to Bergson, something profound is lost: 'real time, regarded as a flux... as the very mobility of being, escapes the hold of scientific knowledge'.[6]

Yet digital audio, despite its high-speed dissection of time into immobile cuts, need not be dismissed as the sterile operation of mechanistic intellect. Its principles are mathematical, its mode is extreme precision. But its properties offer a vast array of potentials for creative use. Time in digital editing is infinitely supple. All of the analogue audio techniques – cutting, fading, mixing, looping, delay, reversing – are honed with greater precision and control in the digital domain. Some techniques, such as vari-speed and delay, benefit from the greater accuracy available, affording more creative scope. Digital editing and delay permit the tiniest fraction of a sound to be repeated indefinitely, as if that sonic material is frozen in time. 'Elastic audio' software, introduced in 2007, enables the user to change an audio file's tempo or timing without altering the pitch. Digital audio editing, then, preserves aspects of a specific sound while allowing other characteristics of the sound to be transformed. That sound may be elongated, bent, repeated, inverted, softened, changed in pitch; it may be altered in its envelope – or its movement through time.

The polyphonic digital sampler, pioneered by the Australian Fairlight company in 1979, and widely available through other manufacturers since the mid-1980s, has added another range of manipulations of sound in time. When played on a sampler keyboard, any sound can be detuned, elongated, looped or sped up, while retaining its basic properties; that is, a sound can be reshaped in time, yet remain recognisable. The sampler plays sounds as if they were musical instruments, inserting their specific sonic profiles into the flow of music in time. A sampling function is included in all contemporary audio editing software, making possible the 'playing' of recorded sounds with altered parameters.

Digital audio presents us with a range of paradoxes. Its high precision encourages non-linear editing, in which material can be retrieved and assembled in any order. Its mathematical nature offers an infinite number of choices in non-destructive editing. It is based on tiny slivers of frozen time, yet it offers inexhaustible means to explore the ambiguities and flux of time. These paradoxes proceed from its central concept, which comprises the greatest paradox. Its stuff is numerical information, yet that stuff is a non-stuff, manipulable to an unprecedented degree. Its binary language is brutally simple, but the ways it invites us to think and create are unfathomably complex.

Visual sound

The first experience of digital editing, for those previously accustomed to analogue sound, has a startling effect. The sense ratios are altered: instead of the tactile/aural configuration of working with magnetic tape, there is a visual/aural alignment. You *see* the sound, a visual representation of the waveform. The Fairlight Computer Music Instrument of 1979 came equipped with a light pen, which the user wielded to shape the image representing the sampled sound. In a digital audio work station, whole sequences of sound are displayed on the screen in visual terms as rectangular shapes, stretches of recorded time waiting to be accessed. On demand, a cursor moves through the sound-as-image: this cursor represents the 'now', the present moment of the displayed sound which we hear as we watch it move across the screen (see Figure 8.1).

There are several consequences of this digital conceptualisation of time. For Bergson, the 'fundamental illusion'[7] about time is to consider it in terms of space: we visualise time, most often represented by a line or an arrow in space. Digital audio, prevalent since the 1980s, has constructed a representation of time in these very terms. Sequences of recorded sound are visualised; they move from left to right across a screen occupying pictorial space within the computer monitor. Yet once again we are presented with paradoxes. Digital editing allows the user to see the whole project as it is being edited – all components, all sequences visualised as stretches of time. Unlike a digital watch, which merely shows the viewer a numerical representation of the present moment – a time sample – the visual display of a digital audio system allows all items of recorded sound to be present. These sounds – their temporal identity transposed to the

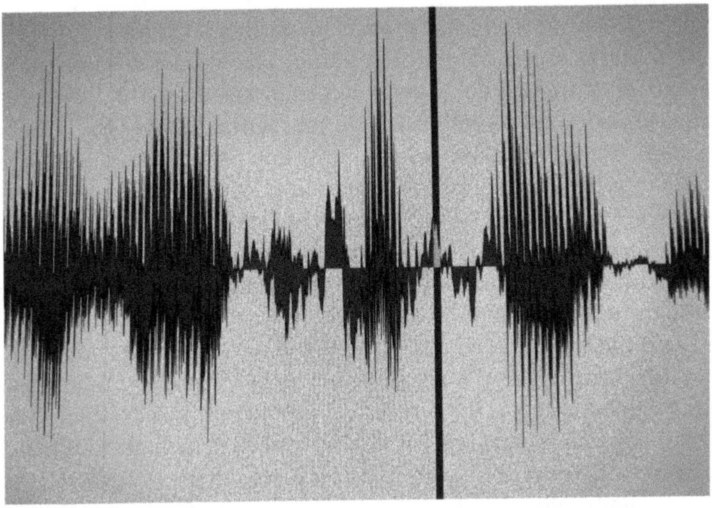

Figure 8.1 An audio waveform. Photo: Marcus Eckermann and Peter Ring

spatial domain – are situated in any order within the visual display, either behind the cursor or ahead of it. The user can see the past and the future (in relation to the cursor, which represents the present); but the positions of these sequences can be re-arranged at will. Time for the user is full of possibilities. It can be reshuffled with ease. It can be intuited in an infinite number of permutations.

R. Murray Schafer coined the term 'schizophonia' in the 1970s to describe the effect of sound-recording technology in splitting a sound from its source, preserving the sound in recorded form.[8] But such a technology also splits a sound from its time: it is 'schizochronic'. In recording a sound, we preserve its flow in time. The recording represents a past sequence of time, which, when played, returns to occupy the present. Any recording is a past waiting to return to the present. The replayed sound is ontologically distinct from the original, since it is a recorded version displaced in both time and space. Its return at a later time is a form of difference: the sound is marked by both the technological intervention and the displacement in time. Incorporating these markings of future difference, the sound once recorded is re-constituted: it is split across time, imbued with the potential of re-emergence in time.

Digital audio technology adds a further dimension to the schizochronic aspect of recorded sound. The visualisation of sound components displays an array of recordings waiting to be retrieved, re-assembled, shuffled with others in the register of time. The cursor – in some systems referred to as 'now-time' – represents a mobile present; jumping forwards and backwards, bringing into the present whichever waveform it lands upon. Time becomes fluid, continually realigned by the mobile cursor. As well, the system permits a rapid shifting of the timescale. The cursor may be observed sweeping through individual sounds, or, from a much larger perspective, the entire project may be represented on the screen. At the moment of looking, the user sees the whole timescale of the project in question. The past and future of the project are apprehended – or rather, a profusion of possible pasts and futures, waiting to be assembled. There is made possible an intuitive grasp of a project's shape in time.

This digital visualisation of sound takes its place in a history of such technological developments. Edison's phonograph of 1877 was intended by its inventor primarily for 'letter-writing and all kinds of dictation';[9] it was, as its name suggested, designed for 'sound-writing'. Both Jacques Attali and Marshall McLuhan have commented on the birth of sound-recording technology in the nineteenth century, in an age culturally dominated by the visual medium of print. Ironically, the phonograph, along with its predecessors and rivals such as the phonautograph and gramophone (*gramma* = letters), shared the aim of printing: 'to transform sound into writing'.[10] So dominant was the concept of mechanical writing in 'the Gutenberg era with its smooth, uniform lines of type and organisation'[11] that the inventors could not conceive of their sound-recording devices except in terms of inscription. This predisposition to the written word was also, of course, a bias towards the visual; for McLuhan its influence was not supplanted until the 1950s with the emergence of hi-fi and the availability of the tape recorder. The emphases of this electric era were tactility and acoustic space, representatives of an episteme vastly different from that of only a half-century before, in which the 'graphophone' needle was conceptualised as a type of pen to write sound.[12]

The age of digital sound, commencing three decades later, problematises these relations in a new way. Digital audio contributes to a changed conceptualisation and experience of sound. Certainly, the visual display of sound constructs the user's experience in ways absorbed from earlier media, including print. The role of vision is privileged. We read, or perhaps scan, the information from left to right (in the same way that Western music notation borrowed this convention from print). We are encouraged to consider sound items in linear terms, as stretches of time, visually represented. Yet there are complicating factors. The experience of digital audio is a visual-aural synchronisation: the visual is privileged, but not pre-eminent. The linear movement of items in the visual display is subsumed into a greater whole within digital editing; non-linear editing, as already discussed, incorporates movement from right to left, as well as the testing of unlimited versions of the edited sequence.

In addition, the visual display of multitrack mixing adds the element of vertical movement, as sections from one track may be moved up or down to another track. Digital mixing represents the multilinear in visual terms; it is a visualisation of the 'all-at-onceness' of multitrack mixing, in which complex combinations of sounds can be grasped in visual terms – as they are heard, or, indeed, before they are heard. There is a complicated and rapid interplay between the visual and aural senses in digital editing/mixing. The visual display is not merely a representation of sound in spatial terms; it has a temporal dimension as well. It affords an intimation of elaborate sound constructions before they are heard. The eye, in combination with the ear, builds complex layers of sound in time.

The digital age of convergence renders all information – audio, video, text – into streams of numerical data. Multimedia formats meld previously discrete media forms into new amalgams. Technological sound now has a visual component; it is capable of both great precision and great subtlety. Digital audio incorporates contrasting facets of time. Founded on the rationalist principles of time dissection, it nevertheless affords enormous scope for the creative manipulation of sound in time. It renders time both as constructed by the intellect, and as glimpsed by intuition. In making possible the play of sound in time, including the intuition of multiple possibilities unfolding in time, it suggests something of the 'infinite multiplicity of becomings' exhorted by Bergson.

9
Capsules of Time and Space: Video and Performance Art

We may expect some aspects of the contemporary experience of time and space to be reflected in recent video art and performance artworks. These two relatively young art forms, both having emerged in the 1960s, have often featured the construction of time and space as a core characteristic of specific works. These artworks have at times reflected the complexity of hybrid space, in which the virtual overlaps with physical space. Time has been assayed in myriad ways: the subjective experience of time has been weighed against the chronological charting of time; brief intensities of time have been contrasted with long, demanding durations. Video artists, exploiting the possibilities of digital video and computer editing, have created video works that manipulate time and build contained and particular versions of space. Performance art in many ways has been concerned with time: the real time of live performance, mediated time, extended duration, the psychological experience of time. Both art forms have produced capsules of spacetime that reveal facets of the contemporary experience and conceptualisation of space and time.

Across the universe: Literature and film

A concern with time and space is also evident in recent narrative-based works in literature and cinema. Time features almost as a character in some of the novels published since 2008, including Julian Barnes' *The Sense of an Ending* (2010) and Per Petterson's *I Curse the River of Time* (originally published in Norwegian in 2008, English translation 2010). The English novelist Barnes inserts a meditation on time into the first page of *The Sense of an Ending*:

> We live in time – it holds us and moulds us – but I've never felt I understood it very well. And I'm not referring to theories about how it bends and doubles back, or may exist elsewhere in parallel versions. No, I mean, ordinary, everyday time, which clocks and watches assure us passes regularly... And yet it takes only the smallest pleasure or pain to teach us time's malleability. Some emotions speed it up, others slow it down; occasionally it seems to go missing – until the eventual point when it really does go missing, never to return.[1]

The writer and critic Colm Tóibín noted that *The Sense of an Ending* 'allows paragraphs about time and memory to dominate rather than paragraphs about character and motive'. Time in this novel often appears to be 'a great deal more solid than we think, or than we imagined'.[2]

The narrator's reflections on time and its malleability depending on the individual's psychological state presage the theme of the novel, which also explores the construction of the past in memory. The conceptualisation of time in twenty-first-century physics – the 'parallel versions' of time postulated in multiverse theory – is cited as a theoretical backdrop, suggesting a complexity in the possibilities of time in contemporary thought. But it is time as perceived by consciousness – a well-travelled theme of the novel at least since Proust – that principally occupies the narrator of *The Sense of an Ending*. His memory of a past shared with others is disrupted by new evidence, so that his version of the past is overturned: the central symbol in the novel is of a river suddenly – and in an isolated spot – running backwards. Memory is re-made and time – or at least memory of past events – is re-constructed.

The Norwegian novelist Per Petterson has a similar fascination with memory and the shifting construction of the past. *I Curse the River of Time* (the title comes from a poem by Chairman Mao) shifts through time-periods, as the narrator remembers events in 1989 – as communism in Europe is collapsing – and as his mother is dying. The characters' concern in 1989 is with the narrator's earlier decision, as an idealist Marxist youth, to abandon his university studies to take up work in a factory – a decision satisfying his need to be proletarian, but one that dismayed his mother, who had worked for decades to allow her son a college education.

In this novel, the past is continually infiltrating the present. Childhood memories are vividly recalled, 'with such teeth-grinding intensity that I almost frightened myself', as the narrator states. The critic James Wood has said of this novel in particular, and Petterson's fiction in general, that 'the past ghosts its way back into the present with spectral power'.[3] This Bergsonian persistence of the past in duration is complemented by a range of intensities of time throughout the novel, as if time itself is a multi-dimensioned character. The narrator claims that 'time without warning could catch up with me and run around beneath my skin like tiny electric shocks'. This happened in 1989, as events took the narrator by surprise: 'I had not paid attention . . . Time had passed behind my back and I had not turned to look.'[4] The novel moves within time, along with its narrator's reflections, as memories of key actions – and the regrets that accumulated over decades – play out in his consciousness.

Other recent literary works are notable for their creative investigation of time and space. The German writer W. G. Sebald's innovative works – blends of the novel, essay and memoir – contain powerful meditations on the nature of memory, the means of apprehending the past and the elusive nature of spacetime itself. *Austerlitz* (2001) and *The Rings of Saturn* (1995, English translation 2002) roam across territory and through time, in long passages of text punctuated with photographs. Sebald developed a singular technique for evoking the past in writing that is both lyrical and philosophical. The novels of the English writer David Mitchell feature great sprawling temporalities spread across global space. *Ghostwritten* (1999) moves between Japan, Hong Kong, Mongolia, Russia, England and elsewhere as characters are linked across space through devices including a disembodied spirit that inhabits human hosts. *Cloud Atlas* swings not only across continents but through time, in connected narratives from the nineteenth century to a post-apocalyptic future. Identity itself seems to spiral across time and space in this extraordinary work.

The fluidity and complexity of contemporary spacetime have been reflected in recent cinema, particularly the films of Christopher Nolan – *Memento*, *Inception*, *Interstellar* – and Charlie Kaufman – *Being John Malkovich*, *Synecdoche New York*, *Eternal Sunshine of the*

Spotless Mind, as well as the Wachowskis and Tom Tykwer. These and other recent films display a dazzling array of temporalities and spatial constructions. In some instances, time flows backwards as well as forwards: *Memento* (2000). In Tykwer's *Run Lola Run* (1998) time forks along three paths, as the same narrative sequence plays out – in a model adopted from videogames – in three different versions. In other films, space is presented as a mental architecture – *Inception* (2010), *Eternal Sunshine* (2004) – or hyper-real zone – *Synecdoche New York* (2008); or the complexities of spacetime envisioned by contemporary physics are realised in science fiction – *Interstellar* (2014). The Wachowskis' *Matrix* trilogy (1999–2003) utilised digital manipulation to render a highly elastic version of time and space befitting a virtual reality setting. The Wachowskis and Tykwer combined to direct a cinema adaptation of David Mitchell's spacetime saga *Cloud Atlas* (2012). These recent explorations in cinema probe and extend the imagined reaches of time and space.

Time-space capsules: Video art

Video art emerged in the late 1960s, following the development in 1965 of portable video cameras such as the Sony Portapak, made available to members of the public at prices much cheaper than film cameras. By 1968, exhibitions of video art had been staged across Europe, South America, Japan and the US.[5] The new art form was born at a time when the boundaries between art practices had become blurred, as Michael Rush has pointed out in his history of video art. Performance, dance, music, sculpture, writing, body art, conceptual art: all these forms and practices could be melded in the new electronic form of video. The history of film and the conventions of television were also reference points for the emergent video artists.

Video was an 'art of time', but that art was not necessarily tied to narrative, as was the case of cinema. Rush observes that in each stage of its history, video art displayed an interest in time as a medium, and as a function of video itself. Other theorists of video art have noted video's 'liberation' of time from the formal constraints of narrative. Non-narrative video works take the nervous edge off narrative space, in the words of Jean-Paul Fargier, to 'translate the problem of space into a question of time'.[6] Some of the early works were founded on a commitment to 'real time': unedited and unprocessed, capturing time as it was experienced in the present.[7] Works by Vito Acconci, Marina Abramovic and others explored aspects of the body in real time, in a manner connecting these early video works with performance art, itself a newly emergent form.

Later waves of artists, exploiting the possibilities of digital video and computer manipulations of the image, stretched time itself, 'breaking the barriers between past, present and future'. Multi-screen installations can accommodate multiple layers of time, as it is 'experienced in our waking and sleeping states'.[8] Rush summarises the treatment of time in video art as attempts to 'extend, repeat, fast forward, slow down, speed up and stop time'.[9] Video art has realised the 'malleability' of time, as described in Barnes' novel *The Sense of an Ending*: the ways in which it can speed up, slow down and alter its shape.

Space has been an important element in video art as well: the video screen frames a construction of space. The South Korean video artist Nam June Paik's pioneering videos of the 1970s practised a displacement though the video image. In works such as *Video Buddha* (1976), the Buddha statue exists twice, in two different spaces: in the physical space situated in front of the camera, and on the video screen. This closed-circuit installation effects a doubling of the object: it is statue and image of the statue, both existing in real time. The doubling of the object adds a virtual or electronic component, as the representation of the thing exists – for the life of the artwork – on the screen. This influential work by Paik was an early exploration of the hybrid space made possible by mediation and screen technologies. Paik's works became more complex as the technologies of mediation – including the internet – became more complex. In 1995, he mounted a major video installation entitled *Electronic Superhighway*, reflecting the advent of the internet and the increasing virtualisation of space.

Much of the pioneering video art in the 1960s and 1970s was created in an environment rich in conceptual art and a vibrant experimental film culture. In the US, the San Francisco Bay Area was a 'haven for radical art and experimental film and video', as film scholar Steve Anker has noted.[10] The area had embraced a radical media culture since the mid-1940s, building a creative culture full of 'malcontents and revelators', as film historian Steve Seid has observed of the Bay Area.[11] Experimental film-makers including Sydney Peterson and Stan Brakhage embraced plasticity rather than narrative form in their short films; Peterson has suggested that the topography of San Francisco, its spatial dissonances and disorientations, contributed to the avant-garde visual sensibility of experimental film and video made there.

Peterson remarks that the unusually steep hills of San Francisco 'violate that sense of linear perspective' that is 'part of most people's way of viewing things'.[12] The sharp inclines of the city allow 'one to look back at where one has traveled from and to where one is going, conflating past, present and future'. This 'constant play of time and space' encourages adventurous visual artists 'to make daily discoveries', challenging them to make original art.[13] This is the city of *Vertigo,* Hitchcock's most poetic and mysterious film, suffused with the haunting presence of the past and the psychological disorientations of heights. The French video-maker Chris Marker revisited this imagery – and the city – in his *Sans Soleil* (1983): this prolonged essay-film includes meditations on the spirals of time within Hitchcock's film, following the *Vertigo* characters' footsteps through San Francisco.

The unusual topography of the urban area may have been one contributing factor to the focus on space found in many of the conceptual video pieces made in the Bay Area in the 1970s. Steve Seid has called space 'the conceptual frontier' in these works, in which 'the physical arena of the work itself' became a subject of study.[14] Peter d'Agostino's *The Walk Series* (1973–1974) is a construction of the surrounding space through the medium of video: three sites are marked off by the artist using a handheld camera. The landscape is represented on video at ground-level, as the walks play out a square. Seid describes this work as 'a way to exert a sense of presence, aestheticized and otherwise, on an indifferent urban expanse'.[15] The urban terrain is marked – and represented – by video.

A video work from 1975 by Darryl Sapien is a meditation on space that contains a premonition of the virtual. Sapien's *Splitting the Axis* shows the splitting of an enormous pole as filmed by three cameras. A fourth monitor displays a live mix of the three closed-circuit sources. In Seid's view, this mix created 'a compressed record that effaced spatiality in favour of a cinematic rendering'. The fourth monitor displays a mixed transmission composed of video representations of the event: Sapien's prescient work 'saw a coming state in which the body and its virtual foil would part for places unknown'.[16]

Here the artist is functioning as antenna, as early-warning system, anticipating a mediatised world where space is hybrid, built of physical and virtual components. *Splitting the Axis* is a 'pre-ghost of the virtual', Seid observes, using the possibilities of video.[17]

Christian Marclay's *The Clock* (2011) is a large-scale video installation comprising over 10,000 clips from film and television showing the time – as indicated on clocks, watches and other time-pieces, along with slices of narrative action. *The Clock* runs for 24 hours, so that visitors can experience their own section of clock-related viewing: half an hour, an hour or half a day. Although the subject-matter of the work may appear to be clock time, Marclay has professed a weariness for the 'fast edit', claiming that he sought to create with *The Clock* a visual collage 'more connected to the way time flows'.[18] Indeed the experience of time is not uniform for visitors to *The Clock*: at certain points in the 24-hour cycle, the edited clips move faster, are packed with more narrative action and appear more intense than at other times. The early hours of the morning, for example, are less populated with clips and action, and seem to move more slowly.

It is apparent from a lengthy period viewing *The Clock* that its governing principle is rhythm. Thousands of film clips have been stitched together into one gigantic rhythm, occurring in real time. There is a sense of the work in its entirety as a compendium, in which one day is represented by a storehouse of world cinema (it is a film analogue to James Joyce's novel *Ulysses*, which compressed world culture and myth into a single day). The world's activity as represented on film is edited together according to the order of mathematical time: 24 hours in the day, 60 minutes to the hour. But there is a tension throughout the work between this metronomic clock rhythm and the flux of activity depicted in the edited clips. Key moments in time – such as noon – are intensified as the preceding clips build up with rapid shots of watches and clocks, people watching them anxiously or expectantly, creating a heightened rhythm and focus on the time itself rather than on narrative action. After the key moment has passed, and other clips roll on, there is a release of tension, a relaxation post-noon. *The Clock* creates a flow of time, a rolling of duration at varying intensities and speeds, closer to the way time is experienced throughout the day than to the metronomic consistency of actual clock time.

Several artists using the medium of video have been fascinated by the possibility of slowing down time – or at least, slowing down video. The Scottish artist Douglas Gordon slowed Hitchcock's *Psycho* so that its 'running time' lasted a day, in his *24 Hour Psycho* (1993). Gordon chose this film for its familiarity: most visitors to the video installation would have seen this film – but at two frames per second instead of 24, the decelerated film reveals new dimensions. Details missed at normal running speed are detected at slow speed; in addition, the extended screening time stretches linear time for the viewer, creating 'an exceptional impression of duration'.[19] Gordon has described the 'temporal dislocation' intended for the viewer of this work, in that the viewer is 'pulled back into the past in remembering the original, then pushed into the future in anticipation of a preconceived narrative that will never appear fast enough'. Between the pull of the past and the lure of the future is 'the slowly changing present', made out of time with the world around it.[20]

The American artist Bill Viola has specialised in poetic imagery, often with spiritual or religious overtones, often in extreme slow motion. Viola has adopted digital techniques of editing and slowing moving images, as well as technical developments such as the flat LCD screen, which allowed video artworks to be mounted on walls like paintings. From the mid-1990s, his flat screen video works drew on Renaissance painting for inspiration: works such as *The Greeting* (1995) displayed modern versions – in extreme slow motion – of the Renaissance portraits. The *Tristan Project* (2004–2006) included elemental imagery of fire and water, again in extreme slow motion. At times the image in Viola's works moves so slowly that the movement is almost imperceptible: time seems to be suspended. Viola has said that his videos are 'an antidote to the relentless pace of modern life'[21] – a pace caused in part, ironically, by the very digital technologies used by Viola in his video art.

Not all video art is concerned with slowing down time to a standstill. The Australian video artist Shaun Gladwell, in a series of works dating from *Storm Sequence* (2002) has constructed time-space capsules, often of actions performed in urban environments filmed from a static camera. The action is tightly framed by the camera; time operates according to the logic of the piece, generally on a loop. Film theorist Adrian Martin has described these video works as comprising 'blocks of time' and 'blocks of space': Gladwell's video art 're-energises the capacity of audiovisual media to seize space-time blocks'.[22]

The Australian artist Daniel Crooks has developed a 'timeslice' technique in both video works and still images, so that multiple expressions of time are represented within the one image. Crooks has described his process of digital editing of time as deriving from 'new post-camera models of spatio-temporal representation'. As in the digital image *Portrait #2 (Chris)*, time is rendered in 'a more tangible material form' (see Figure 9.1).[23]

The media theorist Kathy Cleland has observed that in Crooks' digital portraits we see 'different temporal moments spatially represented as slices juxtaposed together'. These digital works reveal a range of spatio-temporal views of his portrait subjects in both still and

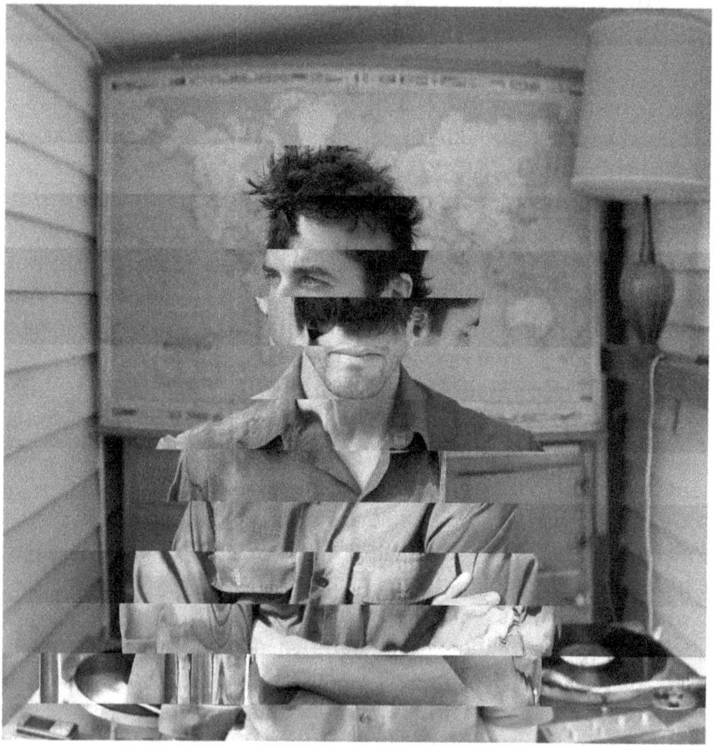

Figure 9.1 Daniel Crooks, *Portrait #2 (Chris)*, 2007. Lambda photographic print, 102 × 102 cm. Courtesy the artist, Anna Schwartz Gallery and featuring Christopher Brown

moving image. There are echoes of both Cubist painting and photomontage in their 'fractured and multi-perspectival aesthetic',[24] but the originality of Crooks' technique is that it incorporates time itself into the picture. In *Portrait #2 (Chris)*, the depicted space is domestic, stable; a map of the world on the wall contains a further representation – or simple mapping – of space. Time is represented, however, in multiple slices; time is mobile within a static image. This is visualised time, or spatialised time, in the construction of time criticised by Bergson as a misrepresentation of duration. Yet, as was noted in consideration of time in digital audio, the fluidity and poetics of time made possible by techniques of digital editing bring us closer, as in Crooks' portraits, to a sense of duration as dynamic and multiple.

Space and timelessness: Performance

Performance art emerged in the 1960s, drawing on precedents in the modernist avant-garde – Futurism, Dada – as well as developments in conceptual art such as composer John Cage's foregrounding of chance in the 1950s. In the early 1960s, performers of George Brecht's 'event pieces' or Allan Kaprow's 'happenings' began to establish an art form based on the live performance of artworks. The German artist Joseph Beuys, seeking to infuse art with shamanic ritual, was another important influence.

For early devotees of performance art, the liveness and singularity of the event was crucial. It was imperative to experience the 'maniacally charged present' of a live performance, in the words of theorist Peggy Phelan: 'Performance's only life is in the present.'[25] A postmodern turn came over performance, however, with the introduction of video documentation, broadcasting of performance events and, in the early twenty-first century, re-enactments of significant performance works from earlier decades.

A central figure in all these events has been the Serbian-born, New York-based performance artist Marina Abramovic. This leading figure in performance art said in 2006 of a performance work: 'What if somebody in New Zealand wants to see it?'[26] The maniacally charged present may have been witnessed by only 20 people in New York; recording and distribution of the performed work ensures a much greater audience. Abramovic also made a major statement in 2005 with a programme of re-enacted works at the Guggenheim in

New York. This event included re-stagings of highly significant works of the 1960s and 1970s by Beuys, Bruce Naumann and others, including Abramovic herself. Performance art had moved into the world of mediation, re-enactment and the circulation of sounds and images across global networks.

Abramovic made a highly publicised series of performances in 2010 at the Museum of Modern Art (MOMA) in New York, entitled *The Artist Is Present.* The artist for these performances was present, sitting in a chair, for 736 hours, while visitors to MOMA took turns sitting opposite her, for as long as they desired. A 2012 feature film, *Marina Abramovic: The Artist Is Present,* was based on these groundbreaking performances. Abramovic arrived at this highly simplified mode of long-duration performance after 40 years of performance art. She described the rationale for this new mode of performance: 'I realize that only long-durational work of art has the serious potential to change the viewer looking at it but also the performer doing it.' She added that 'because the performance is long, it becomes like life itself'.[27]

Abramovic's preparation for long-durational works such as this and *512 Hours,* performed at the Serpentine Galleries in London in 2014, involved a four-day retreat for herself and any other performers, in a purifying ritual she calls 'Cleaning the House'. Participants in this retreat remove watches, time-pieces, phones and digital devices, consume only green tea and water, remain silent, and engage in structured exercises such as observing a single object for hours. This purifying process has echoes of the spiritual practices of shamans and other spiritual leaders in preparation for a psychically charged feat of endurance.

During the performance itself, her goal is the complete engagement of the audience with the work: 'I don't want an audience to spend time with me looking at my work; I want them to be *with* me and forget about time.' A sense of timelessness is a goal of these long-duration performances: 'Open up the space and just that moment of here and now, of nothing, there is no future and there is no past.'[28] As Sophie O'Brien wrote of the *512 Hours* performance work, Abramovic has 'committed within her practice to changing our relationship to time'. She pushes through extended duration to create 'black holes', where the audience 'can experience the depth of time'.[29] These performances create for the audience members a sense of time and its 'intensive magnitude', as Bergson wrote of duration, 'the very flux of the real'.

A similar claim has been made for the performance works of Australian artist Mike Parr. The performance theorist Edward Scheer draws on Bergson's concept of 'real duration' as 'simple flux, a continuity of flowing, a becoming' to describe the experience of time during a Parr performance. Bergson conceptualised duration as a fluid, non-chronological, non-linear form of time; Scheer finds this poetics of time expressed everywhere in Parr's work. The 'aesthetics of duration' is disclosed as a central component in works as different as shock pieces, with their rapid attacks, and slow extended ritual performances. Scheer observes that Parr's performances heighten the perception of time: they slow it down in a kind of meditation, or quicken it in the manner of a traumatic episode. Extremely long ordeal works generate 'a subtle transformation of perception', in which the limits surrounding the work and the self are 'rendered temporal'.[30] Parr himself has claimed: 'Time for me is the meaning of performance art.'[31] Extended-duration performance works stretch out time, allowing us to experience its flux.

Notes

Introduction

1. John Potts, 'Who's Afraid of Technological Determinism: Another Look at Medium Theory' (2008).
2. Wolfgang Ernst, 'Let There Be Irony: Cultural History and Media Archaeology in Parallel Lines' (2005), cited by Jussi Parikka, *What Is Media Archaeology?* p. 83.
3. Friedrich Kittler, 'The Mechanized Philosopher', p. 195.
4. Jack Goody, *The Domestication of the Savage Mind* (1977), p. 128.
5. Walter J. Ong, *Orality and Literacy* (1982), p. 78.
6. Elizabeth Eisenstein, *The Printing Press as an Agent of Change* (1979).
7. Joshua Meyrowitz, *No Sense of Place: the Impact of Electronic Media on Social Behavior* (1985).
8. Pierre Levy, 'Toward Superlanguage', in *International Symposium on Electronic Art 94* catalogue, Helsinki: University of Art & Design, 1994, p. 10.

1 A Brief History of Time and Space

1. Albert Einstein, 'The Problem of Space, Ether and the Field in Physics', *Essays in Science*, cited by Murad D. Akhundov, *Conceptions of Space and Time*, p. 3.
2. Akhundov, *Conceptions of Space and Time*, p. 34.
3. Marie-Louise Von Franz, *Creation Myths*, p. 66.
4. Mircea Eliade, *The Myth of the Eternal Return*, p. 82.
5. Von Franz, *Creation Myths*, p. 62.
6. Ibid., p. 115.
7. Ibid., p. 130.
8. Ibid., p. 192.
9. Cited by Paul Brockelman, *Cosmology and Creation*, p. 29.
10. Akhundov, p. 35, uses the term 'bipresence' in describing the 'supernatural' or 'mystical' appreciation of territory by tribal peoples.
11. Akhundov uses this example of the shaman's cure following a spiritual journey to illustrate 'a world characterized by dual or multiple presence', p. 36.
12. Akhundov, p. 39.
13. Ibid., p. 42.
14. Claude Lévi-Strauss, *The Raw and the Cooked*, p. 16. Lévi-Strauss here likens mythology to music in that both need time 'only in order to deny it'. Myths overcome the 'contradiction between historical, enacted time and a permanent constant'. The translation by John and

Doreen Weightman of Lévi-Strauss' resonant description of myths is 'instruments for the obliteration of time'.

15. Wally Caruana, *Aboriginal Art*, p. 10.
16. Ibid.
17. Eliade, *The Myth of the Eternal Return*, p. 35.
18. Ibid., p. 84.
19. Ibid., p. 79.
20. Ibid., p. 89.
21. Ibid.
22. Richard Cavendish (ed), *Mythology*, p. 25.
23. Ibid., p. 245.
24. Hesiod, *Works and Days*, cited in Cavendish, p. 123.
25. Akhundov, pp. 60–61.
26. The works of all the pre-Socratic philosophers are lost, and are available to us only in the form of quotations, paraphrase and commentary from later authors. Thales cited by Akhundov, p. 61, drawing on Hermann Diels' *Die Fragmente der Vorsokraiter.*
27. Charles Freeman, *The Greek Achievement*, p. 260.
28. Akhundov, p. 64. Heraclitus cited by Clemens Alexandrinus, *Stromateis*.
29. Akhundov interprets Heraclitus' treatise on the 'ever-living fire' in this way, p. 64.
30. Akhundov, p. 66. Heraclitus cited by Plutarch, *De Ei Delphico*.
31. Heraclitus cited by Aetius, *De placitis philosophorum*; Akhundov, p. 65.
32. Akhundov, p. 70.
33. Akhundov, p. 80, citing Aristotle's *Physics* on Plato.
34. Henri-Charles Puech, *Man and Time*, cited by Eliade, *The Myth of the Eternal Return*, p. 89.
35. Ibid.
36. Walter J. Ong, *Orality and Literacy*, p. 82.
37. Ibid., p. 73.
38. Akhundov, p. 87.
39. D. V. Dzhokhadze, *Dialektika Aristotelya*, cited by Akhundov, p. 84.
40. Ibid.
41. 'On Exactitude in Science' is a one-paragraph story by Borges initially published in 1946 and collected in Borges' *A Universal History of Infamy* (1946). It is included in Borges' *Collected Fictions* (1999).
42. Dennis Cosgrove, 'Introduction: Mapping Meaning', in Cosgrove (ed) *Mappings*, pp. 1–2.
43. Brian Harley, 'Deconstructing the Map', cited by John Pickles, *A History of Spaces*, p. 12.
44. Harley and Laxton, *The New Nature of Maps*, cited by Pickles, p. 12.
45. Simon Garfield, *On the Map*, p. 28.
46. Christian Jacob, 'Mapping in the Mind', in Cosgrove (ed) *Mappings*, pp. 40, 41.
47. Ibid., p. 31.
48. Ibid., p. 32.
49. Alessandro Scafi, 'Mapping Eden', in Cosgrove (ed) *Mappings*, p. 54.

50. Jerry Brotton, 'Terrestrial Globalism', in Cosgrove (ed) *Mappings*, p. 75.
51. Scafi, 'Mapping Eden', p. 67.
52. Brotton, 'Terrestrial Globalism', p. 75.
53. Eliade, *The Myth of the Eternal Return*, pp. 106, 132.
54. Felipe Fernández-Armesto, *Ideas that Changed the World*, p. 112.
55. Eliade, *The Myth of the Eternal Return*, p. 104.
56. Ibid., p. 105.
57. Ibid., 106n.
58. Akhundov, p. 93.
59. Ibid., p. 96.
60. J. B. Bury, *The Idea of Progress*, p. 22.
61. Albert Camus, *The Rebel*, p. 189.
62. Hayden White, *The Content of the Form*, pp. 143–144.
63. E. G. Richards, *Mapping Time: The Calendar and Its History*, p. 147.
64. Ibid., p. 239.
65. Ibid., p. 86.
66. Chiara Frugoni, *Inventions of the Middle Ages*, p. 89.
67. Ibid., p. 90.
68. Ibid.
69. Lewis Mumford, *Technics and Civilisation*, p. 14.
70. Ibid.
71. Ibid., p. 15.
72. Bacon's *The Advancement of Learning* (1605), cited by John Pickles, *A History of Spaces*, p. 77.
73. Cosgrove, 'Introduction: Mapping Meaning', p. 19.
74. Ibid., p. 18.
75. Descartes' *Discourse on Method, Optics, Geometry and Meteorology* (1637), cited by Pickles, p. 81.
76. Stephen Hawking (ed) *On the Shoulders of Giants*, p. 726.
77. Newton, *Principia* (1687) in Hawking (ed) *On the Shoulders of Giants*, p. 738.
78. Akhundov, *Conceptions of Space and Time*, p. 116.
79. Ibid., p. 132.
80. The neuroscientist Gerald Edelman, in *Bright Air, Brilliant Fire*, p. 35, argues that Kant's *a priori* categories of time and space are in better accord with modern evidence 'on the neuro-physiological properties of brain cells' than are alternative empiricist philosophical viewpoints. He notes, however, that Kant's position 'is not strictly consistent with developmental studies of how babies gain a sense of space'. The contemporary neuroscience interpretation of time as a construction of the mind is detailed in Chapter 2.
81. Esther Milne, *Letters, Postcards, Email*, p. 40, citing Hannah More (1787).
82. Ibid., p. 33.
83. Ibid., p. 43.
84. Scott McQuire, *Visions of Modernity*, p. 185.
85. Ibid.
86. James Carey, *Communication as Culture*, p. 213, cited by Milne, p. 139.

87. Carey cited in Milne, p. 141.
88. McQuire, p. 198.
89. Ibid.
90. Marshall McLuhan, *Understanding Media*, pp. 11, 12–13, cited by McQuire, p. 185.
91. McQuire, p. 116.
92. E. P. Thompson, 'Time, Work and Industrial Capitalism,' cited in McQuire, p. 120.
93. McQuire, p. 115.
94. Hawking, *On the Shoulders of Giants*, p. 1164.
95. F. T. Marinetti, 'Initial Manifesto of Futurism', in J. C Taylor (ed) *Futurism*, New York, Museum of Modern Art, 1961, p. 124.
96. Bachelard's *La Valeur Inductive de la Relativité* was published in France in 1929; *The New Scientific Spirit* in 1934 and *The Philosophy of No* in 1940.
97. 1908 lecture by Minkowski, cited by Roger Penrose, *The Emperor's New Mind*, p. 250.
98. Penrose, p. 250.
99. Paul Davies, 'Newtonian Time, Einstein, Event Horizons and Black Holes' in Christopher Rawlence (ed) *About Time*, p. 190.
100. Brockelman, *Cosmology and Creation*, p. 56
101. Ibid., p. 55.
102. Ibid.
103. Hawking, *A Brief History of Time*, p. 46.
104. Ibid., p. 47.

2 Theorising Time and Space

1. Letter to William James, May 9, 1908, in Henri Bergson, *Key Writings*, p. 362.
2. Suzanne Guerlac, *Thinking in Time*, p. 20.
3. Letter to William James, May 9, 1908.
4. Bergson, *Time and Free Will*, p. 90.
5. Ibid., p. 106.
6. Bergson, *Matter and Memory*, pp. 193, 194.
7. Ibid., p. 275.
8. Ibid., p. 319.
9. Bergson, *Creative Evolution*, pp. 4, 5.
10. Ibid., p. 5.
11. Ibid., p. 219.
12. Ibid., p. 195.
13. Ibid., pp. 218, 219.
14. Guerlac, *Thinking in Time*, p. 10.
15. Ibid., p. 11.
16. Murad Akhundov, *Conceptions of Space and Time*, p. 153.
17. This limited edition version of *The Sound and the Fury*, with coloured time-scheme supervised by Stephen M. Ross and Noel Polk, was published by the Folio Society in 2012.

18. Bertrand Russell, *The Philosophy of Bergson*, cited by Guerlac, p. 12.
19. Gilles Deleuze, *Bergsonism*, p. 115.
20. Guerlac, p. 177.
21. Mark B. N. Hansen, *New Philosophy for New Media*, cited by Guerlac, p. 193.
22. 'Anri Sala', *The Guidebook*, Documenta (13) Catalogue, 2012, p. 302.
23. Ibid.
24. Giorgio Agamben, *Homo Sacer*, p. 176.
25. Carl Schmitt, *The* Nomos *of the Earth*, p. 48.
26. Ibid., pp. 78, 79.
27. Ibid., p. 49.
28. Ibid., pp. 316, 320.
29. Ibid., p. 355.
30. Mitchell Dean, 'Land and Sea: "In the Beginning All the World Was America" ', in Merewether and Potts (eds) *After the Event*, p. 20.
31. Henri Lefebvre, *The Production of Space*, p. 31.
32. Ibid., p. 53.
33. Ibid.
34. Ibid., p. 55.
35. Ibid., p. 356.
36. Ibid., p. 387.
37. Gaston Bachelard, *The Poetics of Space*, p. xv.
38. Ibid., p. xix.
39. Ibid., p. xxxv.
40. Ibid., p. xxxvi.
41. Ibid., p. 4.
42. Ibid., p. 5.
43. Ibid., p. 7.
44. Fredric Jameson, 'The End of Temporality', p. 2.
45. Jean-François Lyotard, *The Postmodern Condition*, p. xxiv, p. 60.
46. Jameson, 'Postmodernism and Consumer Society', p. 114.
47. Jameson, 'The End of Temporality', p. 2.
48. David Harvey, *The Condition of Postmodernity*.
49. Anthony Giddens, *The Consequences of Modernity* (1990). In her survey of recent theorising of time and space – 'Reconceptualising "Time" and "Space" in the Era of Electronic Media and Communications' (2009) – Panayiota Tsatsou advocates a contemporary incorporation of time-space compression and distanciation into a single thesis, within a framework of mediated time and space.
50. Paul Virilio and Sylvere Lotringer, *Pure War*, p. 34.
51. Virilio, *Speed and Politics*, p. 68.
52. Ibid., p. 133.
53. ' "Is the Author Dead?" An Interview with Paul Virilio by James Der Derian' in *The Virilio Reader*, p. 21.
54. David Eagleman, *Incognito*, p. 53.
55. Ibid., p. 51.
56. Burkhard Bilger, 'The Possiblian', *The New Yorker*, April 25, 2011, p. 60.
57. Ibid., pp. 60–62.

58. Eagleman, p. 52.
59. Lyn Gorman and David McLean, *Media and Society into the 21st Century*, pp. 237–238. Other information on the history of the internet comes from Jill Lepore, 'The Cobweb: Can the Internet Be Archived?' and Graham Meikle and Sherman Young, *Media Convergence*.
60. Manuel Castells et al., *Mobile Communication and Society*, p. 171.
61. Ibid.
62. Castells' 1996 edition of *The Network Society*, cited in Shaun Moores, *Media, Place and Mobility*, p. 85.
63. Moores, p. 86, citing Castells' second edition of *The Network Society*.
64. Castells et al., *Mobile Communication and Society*, p. 171.
65. Ibid.
66. Castells, *The Network Society* (2000), p. 406.
67. Ibid., pp. 468, 467.
68. Heejin Lee and Jonathan Liebenau, 'Time and the Internet at the Turn of the Millennium', p. 44.
69. Tom Chatfield, *How to Thrive in the Digital Age*, p. 30.
70. Panayiota Tsatsou, 'Reconceptualising "Time" and "Space" in the Era of Electronic Media and Communications', pp. 21, 23.
71. Ibid., p. 15.
72. Roger Silverstone, *Media and Morality: On the Rise of the Mediapolis*, cited by Tsatsou, p. 17.
73. Graham Meikle and Sherman Young, *Media Convergence*, p. 154.
74. Ibid., p. 155.
75. Google Instant website, cited by Meikle and Young, p. 163.

3 'No-One Is Where They Are': Virtual Time and Space

1. Rebecca Solnit, 'Diary', *London Review of Books*, August 29, 2013, p. 32.
2. Anna Hart, 'The Cult of the Selfie', *The Sun-Herald, Sunday Life*, January 25, 2015, p. 15.
3. Tom Chatfield, *How to Thrive in the Digital Age*, pp. 28–29, 30.
4. Manuel Castells et al., *Mobile Communication and Society*, p. 1.
5. Rich Ling and Scott W. Campbell, *The Reconstruction of Space and Time: Mobile Communication Practices*, p. 14.
6. Panayiota Tsatsou, 'Reconceptualising "Time" and "Space" in the Era of Electronic Media and Communications', p. 17.
7. Kathleen M. Cumiskey, 'Mobile Symbiosis: A Precursor to Public Risk-Taking Behavior?'
8. Giorgio Agamben, *What Is an Apparatus?* pp. 15–16.
9. Stephen Hutcheon, 'Heads Up: New RAAF Pilots' Smart Helmets Will Cost a Bomb', *Sydney Morning Herald*, February 25, 2015, pp. 8–9.
10. Simon Garfield, *On the Map*, p. 384.
11. Ibid., p. 19.
12. Solnit, 'Diary', p. 33.
13. Rachel Olding, 'Meeting App with New Friends Is Changing the Way We Greet the Future', *Sydney Morning Herald*, October 15, 2011, p. 7.

14. Jurgen Habermas, *The Structural Transformation of the Public Sphere*, p. 176.
15. Ichiyo Habuchi, 'Accelerating Reflexivity' (2005), cited by Maria Miranda, 'Uncertain Spaces: Artists' Exploration of New Socialities in Mediated Public Space', p. 2.
16. Miranda, p. 2.
17. Kenneth J. Gergen, 'Mobile Communication and the Transforming of the Democratic Process', p. 302.
18. Robert Pogue Harrison, *Gardens*, pp. 114, 114–115.
19. Ibid., p. 115.
20. Rich Ling, *New Tech, New Times: How Mobile Communication Is Reshaping Social Cohesion*, p. 187, p. 183
21. Gerard Goggin and Kate Crawford, 'Generation Disconnections: Youth Culture and Mobile Communication', pp. 253, 254.
22. Kazys Varnelis and Anne Friedberg, 'Place: The Networking of Public Space,' p. 15.
23. Ibid., p. 30.
24. Louis Menand, 'Crooner in Rights Spat: Are Copyright Laws Too Strict?' *The New Yorker*, October 20, 2014, p. 85.
25. Scott McQuire, 'The Politics of Public Space in the Media City' (2006), cited by Maria Miranda, 'Uncertain Spaces', p. 1.
26. Anthony Dunne, *Hertzian Tales: Electronic Products, Aesthetic Experience and Critical Design* (1999), cited by Miranda, 'Uncertain Spaces', p. 5.
27. Miranda, p. 7.
28. M. Tuters and Kazys Varnelis, 'Beyond Locative Media' (2006), cited in Miranda, p. 3.
29. Miranda, pp. 3–4.
30. Miranda, p. 4, citing Kanarinka list thread, 2006.
31. Miranda, p. 4.
32. 'Janet Cardiff and George Bures Miller', Documenta (13) Catalogue, p. 334.
33. Ibid.
34. Ibid.
35. Sherry Turkle, *Alone Together*, pp. 202, 11, 208–209.
36. Ibid., p. 155.
37. Ibid., p. 14.
38. Ibid., pp. 268, 294.
39. Ibid., p. 277.
40. Amy McNeilage, 'Curse of the Mobile: Fears for the Art of Conversation', *Sydney Morning Herald*, October 29, 2013, p. 9.
41. Ibid.
42. 'Sounds of Silence a Track Favourite', Letters, *Sydney Morning Herald*, June 7, 2011, p. 11.
43. Derrick Rossignol, 'A New Invention Will Finally Prevent Cellphones from Ruining Concerts' (2014), lists as examples musicians Tyson Ritter and Peter Frampton throwing or destroying cell phones and iPads, and Jack White announcing that he doesn't want cellphones at his performances so everybody can 'just enjoy this with our eyes and ears'.
44. Ibid.

45. Sherry Turkle, *Alone Together*, pp. 279, 17.
46. Matt Wade and Ben Schneiders, 'Under Pressure: Technology Revolutionized Work, Now It's at Home and Sucking Up Our Downtime', *Sydney Morning Herald*, News Review, September 22–23, 2012, p. 1.
47. Ibid.
48. Ibid., p. 4.
49. Ibid.
50. Turkle, *Alone Together*, pp. 16–17.
51. Stephen Hutcheon, 'Watch This Space: The Coming Battle for Your Wrist', *Sydney Morning Herald*, March 6, 2015, p. 13.
52. Max Mason, 'Peppa Pig Future of TV, Says ABC', *Sydney Morning Herald*, March 26, 2015, p. 24.
53. Paul Virilio, *Pure War*, p. 34.
54. Turkle, pp. 171, 172.
55. Peter Munro, 'Switching to Dumb Phones a Smart Move for Some', *Sydney Morning Herald*, March 28–29, 2015, p. 15.
56. Richard Foreman, 'The Pancake People', cited by Nicholas Carr, *The Shallows*, p. 196.
57. Sam Anderson, 'In Praise of Distraction', cited by Carr, p. 140.
58. Turkle, p. 162.
59. Maggie Jackson, *Distracted*, cited by Carr, p. 133.
60. Turkle, p. 163.
61. Daniel J. Levitin, 'Why the Modern World Is Bad for Your Brain', *The Guardian Weekly*, February 6, 2015, p. 27. This article is an extract from Levitin's book *The Organized Mind* (2015).
62. Ibid.
63. Turkle, p. 163.
64. Levitin, p. 27.
65. Ibid., p. 28.
66. Turkle, p. 227.
67. Malinowski cited by Steven Connor, 'Disasters Galore', *London Review of Books*, September 27, 2012, p. 15.
68. Paul Virilio, *Pure War*, p. 35.
69. Matt Wade and Ben Schneiders, 'Under Pressure', p. 4.
70. Solnit, 'Diary', p. 33.
71. Turkle, pp. 273–274.
72. Ibid., p. 274.
73. Peter Munro, 'Switching to Dumb Phones', p. 15.
74. Rachel Olding, 'Gen Y Finds Life Goes on Without Gadgets', *Sydney Morning Herald*, May 7–8, 2011, p. 5.

4 Space and Displacement in Contemporary Art

1. Rex Butler, 'The World Is Not Enough' in Charles Merewether and John Potts (eds) *After the Event: New Perspectives on Art History*, p. 58
2. Nicholas Bourriaud, *The Radicant*, p. 43.

3. Ibid.
4. Ibid., p. 51.
5. Ibid., p. 186.
6. Ibid., p. 188.
7. Charles Baudelaire, *The Painter of Modern Life and Other Essays*, pp. 6–12, cited in Bourriaud, p. 92.
8. Wayne Tunnicliffe, 'Francis Alÿs' in Wayne Tunnicliffe (ed) *John Kaldor Family Collection: Art Gallery of New South Wales*, p. 170.
9. Ibid., p. 170.
10. Alÿs quoted in Tunnicliffe, p. 170.
11. Bourriaud, *The Radicant*, p. 132.
12. Ibid., p. 138.
13. Ibid., p. 140.
14. Butler, 'The World Is Not Enough', p. 58.
15. Carolyn Christov-Bakargiev, 'Introduction', *The Guidebook*, Documenta (13) Catalogue, 2012, p. 7.
16. Ibid.
17. Nicholas Bourriaud, *Postproduction: Culture as Screenplay*, p. 19.
18. Ezra Pound, *ABC of Reading*, p. 73.
19. Marshall McLuhan, *Understanding Media*, p. xi.
20. Mitchell Dean, 'Land and Sea: "In the Beginning All the World Was America"' in Charles Merewether and John Potts (eds) *After the Event: New Perspectives on Art History*, p. 20.
21. Ibid.
22. Bill Fontana, 'The Environment as a Musical Resource' (2000) at http://resoundings.org/Pages/musical%20resource.html. Accessed June 10, 2013.
23. Butler, 'The World Is Not Enough', p. 63.
24. Chia Chi Jason Wang, 'Chen Chieh-jen' in Charles Merewether (ed) *Zones of Contact: 2006 Biennale of Sydney Catalogue*, Sydney: Biennale of Sydney, 2006, p. 96.
25. Ibid.
26. Ibid.
27. *Never Forever*, Exhibition notes, *Runway* #26, www.runway.org.au, Accessed November 18, 2014.
28. Terry Smith, 'Art of Dissent', *The Monthly*, June 2011, p. 58.
29. Ibid.
30. Giorgio Agamben, *Homo Sacer: Sovereign Power and Bare Life*, p. 123.
31. Ibid.
32. Ibid., pp. 168–169.
33. Ibid., p. 166.
34. Ibid., p. 117.
35. Dean, 'Land and Sea: "In the Beginning All the World Was America"', p. 32.
36. Andrew Murphie and John Potts, *Culture and Technology*, p. 171.
37. Agamben, *Homo Sacer*, pp. 175–176.
38. Edward Scheer, *The Infinity Machine: Mike Parr's Performance Art 1971–2005*, p. 129.

39. 'Theaster Gates', *The Guidebook*, Documenta (13) Catalogue, 2012, p. 430.
40. Butler, 'The World Is Not Enough', p. 58.
41. Mark Spiegler, 'When Human Beings Are the Canvas' in *Artnews* June 2003, at http://www.artnews.com/issues/article.asp?art_id=1335. Accessed June 6, 2012.
42. Dean, 'Land and Sea: "In the Beginning All the World Was America"', p. 28.
43. Colin Ripley, 'Hearing Places: Sound in Architectural Thought and Practice', in Bandt et al. (eds) *Hearing Places: Sound, Place, Time and Culture*, p. 87.
44. Ibid.
45. Ibid.
46. Yi-Fu Tuan, *Space and Place: The Perspective of Experience*, p. 3. A summary of theoretical approaches to place within human geography is provided by Tim Cresswell in *Place: A Short Introduction*, Malden: Blackwell, 2004.
47. Lucy Lippard, *The Lure of the Local: Senses of Place in a Multicentred Society*, p. 7.
48. Rebecca Solnit, *Infinite City: A San Francisco Atlas*, p. vii.
49. Ibid., pp. vii, 2.
50. Iain Sinclair, *London Orbital*, London: Granta, 2002.
51. Ros Bandt, Michelle Duffy and Dolly MacKinnon, 'Introduction', in Bandt et al. (eds) *Hearing Places*, p. 1.
52. R. Murray Schafer, *The Tuning of the World*. This book articulated the research conducted by the Vancouver-based World Soundscape Project (WSP), directed by Schafer in the early 1970s. WSP compositions by Schafer, Hildegard Westerkamp and other composers used field recordings to portray acoustic environments.
53. 'Janet Cardiff & George Bures Miller', *The Guidebook*, Documenta (13) Catalogue, 2012, p. 334.

5 The Big Now and the Faraway Then: Present, Past and Future in Contemporary Culture

1. Fredric Jameson, 'Postmodernism and Consumer Society', p. 125.
2. John Potts, 'The Event and its Echoes' in Merewether, Charles and Potts, John (eds) *After the Event: New Perspectives on Art History*, p. 192.
3. Fernand Braudel, *On History*, p. 27.
4. Ibid., pp. 67, 209.
5. Jill Lepore, 'The Cobweb: Can the Internet Be Archived?', *The New Yorker*, January 26, 2015, p. 36.
6. Ibid., p. 38.
7. Ibid., p. 36.
8. Evgeny Morozov, 'What Do We Think About? Who Gets to Do the Thinking?' in Brockman, John (ed) *Is the Internet Changing the Way You Think?* p. 228.
9. Ibid.

10. Ibid., p. 229.
11. Lev Manovich, 'Future Fictions' in *Frieze* No. 156, June–August 2013, p. 199.
12. Ibid.
13. Ibid.
14. George Packer discusses this contemporary ideology in 'Change the World: Silicon Valley Transfers Its Slogans – and Its Money – to the Realm of Politics', *The New Yorker*, May 27, 2013.
15. Cited in Cohen, M. J. and Major, John (eds) *History in Quotations*, London: Cassell, 2004, p. xxv.
16. Of the many studies of the history of progress and its genesis in the Enlightenment, the classic work is J. B. Bury, *The Idea of Progress: An Inquiry into Its Origin and Growth*, 1955 [1920].
17. The Marquis de Condorcet's *An Outline of the Progress of the Human Spirit*, published in 1794, predicted an imminent stage in history in which humanity would achieve perfection through the application of science and reason.
18. Jurgen Habermas, 'Modernity: An Incomplete Project', p. 5.
19. F. T. Marinetti, *Selected Writings*, p. 82.
20. Marinetti, Initial Manifesto of Futurism, in J. C. Taylor (ed) *Futurism*, p. 124.
21. McLuhan develops this idea in *The Medium Is the Massage* (with Quentin Fiore) where he quotes this Dylan lyric on p. 105.
22. Marshall McLuhan, 'The Medium Is the Message' in *Understanding Media*, p. 26.
23. Paul Levinson makes this argument in *Digital McLuhan*.
24. Emily Nussbaum argues that differing attitudes to privacy define a new generation gap in 'Kids, the Internet and the End of Privacy', *Weekend Australian Magazine*, March 24–25, 2007, pp. 23–27.
25. Marinetti, *Selected Writings*, p. 106.
26. Patrick Parrinder, *Shadows of the Future: H. G. Wells, Science Fiction and Prophecy*, p. 136, referring to Wells' *World Brain* (1938).
27. Pierre Levy, *Collective Intelligence: Mankind's Emerging World in Cyberspace*.
28. Quoted in Ken Auletta, 'Publish or Perish: Can the iPad Topple the Kindle and Save the Book Business?' *The New Yorker*, April 26, 2010, p. 26.
29. I parodically characterise the rhetoric of digital prophets in this manner in the essay 'Book Doomsday: The March of Progress and the Fate of the Book' in John Potts (ed) *The Future of Writing*.
30. Simon Reynolds, *Retromania: Pop Culture's Addiction to Its Own Past*, p. 425.
31. Marinetti, *Selected Writings*, p. 95.
32. Ben Knox Miller of The Low Anthem, interviewed in Jaan Uhelszki, 'Children of the Evolution', *Uncut* magazine, March 2010, p. 36.
33. Friedrich Nietzsche, *Untimely Meditations*, p. 115.
34. J. C. Beaglehole (ed), *The Endeavour Journal of Joseph Banks*, p. 145.
35. Christopher Lasch, *The True and Only Heaven: Progress and Its Critics*, p. 83.
36. A range of perspectives on nostalgia and media is provided in the edited collection *Media and Nostalgia*, edited by Katharina Niemeyer, 2014.

37. Charles Merewether and John Potts, 'Introduction' in *After the Event: New Perspectives on Art History*, p. 6.
38. Ibid.
39. Steven Rose, *The Making of Memory*, London: Bantam Press, 1991, p. 91.
40. Ibid.
41. Charles Fernyhough in *Pieces of Light: The New Science of Memory*, p. 14, lists memory researchers subscribing to the reconstructive view of remembering as including Daniel Schachter, Elizabeth Loftus, Endel Tulving, Donna Rose Addis, Antonio Damasio and Martin Conway.
42. Ibid., p. 6.
43. Ibid., p. 14.
44. Igor Stravinsky, *Poetics of Music in the Form of Six Lessons*, Cambridge: Harvard University Press, 1970, p. 57.
45. William Faulkner, *Requiem For a Nun*, Act 1 Scene 3.
46. Michael Prodger, 'Whatever Else Is Going On, History Is Always Present', *The Guardian Weekly*, October 10, 2014, p. 41.
47. Rosalind E. Krauss, 'Dialectic of Enlightenment: William Kentridge's Norton lectures', *Art Forum*, Summer 2012, p. 122.
48. William Kentridge, 'Felix in Exile: Geography of Memory' (extract) in *William Kentridge* Catalogue, London: Phaidon, 2003, p. 122.
49. Alan Gilbert, '"It's Unclear Why..."/Walid Raad and the Writing of History' in *Zones of Contact: 2006 Biennale of Sydney Catalogue*, Sydney: Biennale of Sydney, 2006, p. 80.
50. Tamara Trodd, 'Thomas Demand, Jeff Wall and Sherrie Levine: Deforming "Pictures"', *Art History*, Vol. 32 No. 5, 2009, p. 969.
51. Sophie Forbat, 'Thomas Demand and the CTA' in *The Dailies* catalogue, Sydney: MACK and Kaldor Public Art Projects, 2012, p. 5.
52. Ibid., p. 7.

6 Public Intimacy: The Shrinking Space of Privacy

1. *Coexisting*, commissioned and performed by the artist Clark Beaumont for Kaldor Public Art Project 27: *13 Rooms*, curated by Klaus Biesenbach and Hans Ulrich Obrist, at Pier 2/3 in Sydney, April 11–21, 2013.
2. Robert Ellis Smith, *Ben Franklin's Website* (2004), p. 8.
3. Jordan Robertson, 'NSA Spying Shows Perils of Apps', *Sydney Morning Herald*, January 31, 2014, p. 16.
4. Julia Fioretti, 'Google Starts to Block Search Results After Privacy Ruling', *Sydney Morning Herald*, June 28–29, 2014, p. 44.
5. *New York Times* editorial cited by Jeffrey Toobin, 'The Solace of Oblivion', *The New Yorker*, September 29, 2014, p. 28.
6. Karen Kissane, 'At Last, A Reporter's Insight into Life Under the "Criminal-in-Chief"', *Sydney Morning Herald*, December 1, 2011, p. 13.
7. McNealy and Schmidt quoted by Sherry Turkle, *Alone Together*, p. 256.
8. Andrew Stevenson, 'Cyberspace: It's the New Toilet Wall', *Sydney Morning Herald*, July 21–22, 2007, p. 30.

9. 'Privacy No Longer a Social Norm, Says Facebook Founder', *The Guardian*, January 11, 2010, at http://www.theguardian.com/technology/2010/jan/11/facebook-privacy.
10. Zuckerberg quoted by Jeff Jarvis, 'Privacy, Publicness and the Web: A Manifesto' at http://ieeexplore.ieee.org/stamp/stamp.jsp?arnumber=5779789. Accessed March 27, 2015.
11. Davies quoted in Murad Ahmed, 'Zuckerberg's Revamp Plans Causing Friction', *The Australian*, November 16, 2011, p. 12.
12. Emily Nussbaum, 'Kids, the Internet and the End of Privacy', *The Weekend Australian Magazine*, March 24–25, 2007, p. 24.
13. Sherry Turkle, *Alone Together*, p. 254.
14. Georges Duby (ed) *A History of Private Life*, five volumes (1988).
15. Witold Rybczynnski, *Home: A Short History of an Idea*, p. 54.
16. Ibid., p. 51.
17. Phillippe Ariès, *Centuries of Childhood: A Social History of Family Life*, p. 398.
18. *Monthly Review* of 1782 quoted by Claire Brock, 'Rousseauvian Remains', p. 143.
19. This description of citizen journalists, bloggers and social media users was made by media scholar and writer Jay Rosen on his blog PressThink in 2006.
20. Matt Buchanan and Scott Ellis, 'CCTV Footage Helps Case of Missing Pup', *Sydney Morning Herald*, October 17, 2012, p. 13.

7 Photography 2.0: Photos on the Loose

1. Boris Groys, 'Two Cures: Making Art Visible', p. 10.
2. Ibid.
3. Ibid.
4. Anna Hart, 'The Cult of the Selfie', *The Sun-Herald, Sunday Life*, January 25, 2015, p. 15.
5. Laurent Gaveau quoted in Nick Miller, 'Google Explores Future of Art', *Sydney Morning Herald*, March 14–15, 2015, p. 28.
6. Lina Ben Mhenni, 'The Tunisian Revolution Is Not Over', *Re-Public: Re-Imagining Democracy* (English version) at http://www.re-public.gr/en/?p=4747. Accessed March 15, 2012.
7. Ruth Pollard, 'Voice Behind a Revolution', *Sydney Morning Herald*, January, 21–22, 2012, p. 10.
8. Lina Ben Mhenni, 'The Tunisian Revolution Is Not Over'.
9. Ben Grubb, 'Selfies Get the Short End of the Stick as "Ultimate Narcissism"', *Sydney Morning Herald*, February 25, 2015, p. 3.
10. Sherry Turkle, *Alone Together*, p. 191.
11. Anna Hart, 'The Cult of the Selfie', p. 15.
12. Ibid., p. 192.
13. Ibid., p. 175.
14. Deirdre McKay, 'On the Face of Facebook: Historical Images and Personhood in Filipino Social Networking', p. 480.

15. Ibid., p. 489.
16. S. Dalsgaard, 'Facework on Facebook' (2008), cited by McKay, p. 486.
17. McKay, p. 486.
18. Ibid., p. 487.

8 Schizochronia: Time in Digital Sound

1. Pierre Levy, 'Toward Superlanguage', in *International Symposium on Electronic Art 94* catalogue, Helsinki: University of Art & Design, 1994, p. 10. This is an updating of the proposal made in 1977 by Jack Goody that writing is an 'intellectual technology' that creates a 'different cognitive potentiality' (*The Domestication of the Savage Mind*, p. 128).
2. Ken C. Pohlmann, *Principles of Digital Audio*, Indiana: SAMS, 1992, p. 40.
3. Henri Bergson, *Creative Evolution*, p. 322.
4. Ibid., p. 321.
5. Gilles Deleuze, *Cinema 1: The Movement-Image*, p. xiv.
6. Bergson, *Creative Evolution*, p. 355.
7. Genevieve Lloyd, *Being in Time*, p. 101.
8. R. Murray Schafer, *The Tuning of the World*, Philadelphia: University of Pennsylvania Press, 1980, p. 90.
9. Jacques Attali, *Noise: The Political Economy of Music*, p. 93.
10. Ibid., p. 91.
11. Marshall McLuhan, *Understanding Media*, p. 297.
12. Ibid., p. 293.

9 Capsules of Time and Space: Video and Performance Art

1. Julian Barnes, *The Sense of an Ending*, p. 3.
2. Colm Tóibín, 'Going Beyond the Limits', *New York Review of Books*, May 10, 2012, p. 11.
3. James Wood, 'Late and Soon: The Novels of Per Petterson', *The New Yorker*, December 10, 2012, p. 81.
4. These sentences from *I Curse the River of Time* are quoted by Wood, pp. 82–83.
5. Michael Rush, *Video Art*, p. 7.
6. Jean-Paul Fargier, 'The Hidden Side of the Moon' (1984) quoted by Adrian Martin, 'Chopper: An Essay Around and About Shaun Gladwell', 2011, p. 32.
7. Ibid., p. 10.
8. Ibid.
9. Ibid., p. 8.
10. Steve Anker, 'A Haven for Radical Art and Experimental Film and Video', in Anker et al. (eds) *Radical Light*, 2010, p. 8.
11. Steve Seid, 'Introduction' in Anker et al. (eds) *Radical Light*, 2010, p. 13.

12. Sydney Peterson quoted by Anker, p. 8.
13. Anker, *Radical Light*, p. 8.
14. Steve Seid, 'Thinking Inside the Box: Conceptual Video in the Seventies' in Anker et al. (eds) *Radical Light*, 2010, p. 168.
15. Ibid.
16. Ibid.
17. Steve Seid interviewed by the author, at Pacific Film Archive, Berkeley, California, November 19, 2013.
18. Christian Marclay cited by Helen Powell, *Stop the Clocks!*, p. 16.
19. A. Barak quoted by Helen Powell, *Stop the Clocks!*, p. 62.
20. Douglas Gordon (1996) quoted by Powell, p. 62.
21. Andrew Taylor, 'Interview: Bill Viola', *Sydney Morning Herald, Spectrum*, December 13–14, 2014, p. 5.
22. Adrian Martin, 'Chopper: An Essay Around and About Shaun Gladwell', 2011, p. 32.
23. 'Portraiture in a Digital Age: Daniel Crooks', at http://www.dlux.org.au/face2face/crooks.htm. Accessed March 30, 2015.
24. Kathy Cleland, 'The New Face of Portraiture in a Digital Age' at http://www.dlux.org.au/face2face/Cleland_essay.htm. Accessed March 30, 2015.
25. Peggy Phelan, *Unmarked: The Politics of Performance*, p. 146.
26. Marina Abramovic (2006) quoted by Edward Scheer, '"What If Somebody in New Zealand Wants to See It?" Performance Art's Cover Versions', p. 224.
27. Mark Juddery, 'The Method in Marina's Magnetism', *Sydney Morning Herald*, June 5, 2012, p. 10.
28. 'Marina Abramovic on Performance Art', *Marina Abramovic: The Artist Is Present*, Catalogue, New York: Museum of Modern Art, 2010.
29. Sophie O'Brien, 'A Resonant Emptiness', *512 Hours*, Catalogue, London: Serpentine Galleries, 2014, p. 29.
30. Edward Scheer, *The Infinity Machine*, p. 11.
31. Mike Parr (1979) quoted by Scheer, p. 1.

Bibliography

Agamben, Giorgio, *Homo Sacer: Sovereign Power and Bare Life*, trans. D. Heller-Roazen, Stanford: Stanford University Press, 1998.

Agamben, Giorgio, *What Is an Apparatus? And Other Essays*, trans. David Kishik and Stefan Petadella, Stanford: Stanford University Press, 2009.

Akhundov, Murad D., *Conceptions of Space and Time: Sources, Evolution, Directions*, trans. Charles Rougle, Cambridge, MA: MIT Press, 1986.

Anderson, Benedict, *Imagined Communities: Reflections on the Origin and Spread of Nationalism*, London: Verso, 1991.

Anker, Steve, 'A Haven for Radical Art and Experimental Film and Video', in Anker, Steve, Gevitz, Kathy and Seid, Steve (eds) *Radical Light: Alternative Film and Video in the San Francisco Bay Area, 1945–2000*, Berkeley: University of California Press, 2010.

Anker, Steve, Gevitz, Kathy and Seid, Steve (eds) *Radical Light: Alternative Film and Video in the San Francisco Bay Area, 1945–2000*, Berkeley: University of California Press, 2010.

Ariès, Phillippe, *Centuries of Childhood: A Social History of Family Life*, trans. Robert Baldick, New York: Alfred A. Knopf, 1962.

Attali, Jacques, *Noise: The Political Economy of Music*, Minneapolis: University of Minnesota Press.

Augé, Marc, *Non-Places: Introduction to an Anthropology of Supermodernity*, London: Verso, 1995.

Bachelard, Gaston, *The Poetics of Space*, trans. Maria Jolas, Boston: Beacon Press, 1994 [1958].

Bandt, Ros, Duffy, Michelle and MacKinnon, Dolly (eds) *Hearing Places: Sound, Place, Time and Culture*, Newcastle: Cambridge Scholars Publishing, 2007.

Barnes, Julian, *The Sense of an Ending*, London: Jonathan Cape, 2011.

Barnes, Lucinda (ed) *Measure of Time*, Berkeley: University of California, 2007.

Beaglehole, J. C. (ed), *The Endeavour Journal of Joseph Banks 1768–1771* Vol. II, Sydney: Public Library of NSW/Angus and Robertson, 1962.

Benjamin, Walter, *Illuminations*, in Arendt, Hannah (ed) trans. Harry Zohn, London: Fontana, 1982.

Bergson, Henri, *Matter and Memory*, trans. Nancy Margaret Paul and W. Scott Palmer, London: George Allen & Unwin, 1919 [1896].

Bergson, Henri, *Time and Free Will*, trans. F. L. Pogson, London: George Allen & Unwin, 1910 [1888].

Borges, Jorge Luis, *Collected Fictions*, trans. Andrew Hurley, New York: Penguin, 1999.

Bergson, Henri, *Key Writings*, in Keith, Ansell Pearson and Mullarkey, John (eds) New York: Continuum, 2002.

Bergson, Henri, *Creative Evolution*, trans. Arthur Mitchell, Basingstoke: Palgrave Macmillan, 2007 [1907].

Bourriaud, Nicholas, *Postproduction: Culture as Screenplay*, New York: Lukas & Sternberg, 2002.

Bourriaud, Nicholas, *The Radicant*, New York: Lukas & Sternberg, 2009.

Braudel, Fernand, *On History*, trans. Sarah Matthews, Chicago: University of Chicago Press, 1980.

Brock, Claire, 'Rousseauvian Remains', *History Workshop Journal* (Spring) Vol. 55 No. 1, pp. 134–151.

Brockelman, Paul, *Cosmology and Creation: The Spiritual Significance of Contemporary Cosmology*, Oxford: Oxford University Press, 1999.

Brockman, John (ed) *Is the Internet Changing the Way You Think?* New York: Harper, 2012.

Bury, J. B., *The Idea of Progress: An Inquiry into Its Origin and Growth*, Dover: New York, 1955 [1920].

Butler, Rex, 'The World Is Not Enough' in Merewether, Charles and Potts, John (eds) *After the Event: New Perspectives on Art History*, Manchester: Manchester University Press, 2010.

Camus, Albert, *The Rebel*, London: Penguin, 1971 [1951].

Carr, Nicholas, *The Shallows: How the Internet Is Changing the Way We Think, Read and Remember*, London: Atlantic Books, 2010.

Caruana, Wally, *Aboriginal Art*, London: Thames & Hudson, 2003.

Carey, James W., *Communication as Culture: Essays on Media and Society*, Boston: Unwin, 1989.

Castells, Manuel, *The Rise of the Network Society*, Second Edition, Oxford: Blackwell, 2000.

Castells, Manuel, Fernández-Ardèvol, Mireira, Linchuan Qiu, Jack and Sey, Araba, *Mobile Communication and Society: A Global Perspective*, Cambridge, MA: MIT Press, 2007.

Cavendish, Richard (ed) *Mythology: An Illustrated Encyclopedia*, London: Orbis Publishing, 1980.

Chatfield, Tom, *How to Thrive in the Digital Age*, London: Macmillan, 2012.

Cohen, M. J. and Major, John (eds) *History in Quotations*, London: Cassell, 2004.

Cosgrove, Dennis (ed) *Mappings*, London: Reaktion Books, 1999.

Cresswell, Tim, *Place: A Short Introduction*, Malden: Blackwell, 2004.

Crowley, Des and Williamson, Clare, *The World of the Book*, Melbourne: Miegunyah Press, 2007.

Cumiskey, Kathleen M., 'Mobile Symbiosis: A Precursor to Public Risk-Taking Behavior?' in Ling, Rich and Campbell, Scott W. (eds) *Mobile Communication: Bringing Us Together and Tearing Us Apart*, New Brunswick: Transaction, 2011.

Davies, Paul, 'Newtonian Time, Einstein, Event Horizons and Black Holes' in Rawlence, Christopher (ed) *About Time*, London: Jonathan Cape, 1985.

Dean, Mitchell, 'Land and Sea: "In the Beginning All the World Was America"' in Merewether, Charles and Potts, John (eds) *After the Event: New Perspectives on Art History*, Manchester: Manchester University Press, 2010.

Deleuze, Gilles, *Cinema 1: The Movement-Image*, trans. Hugh Tomlinson and Barbara Habberjam, Minneapolis: University of Minnesota Press, 1986.

Deleuze, Gilles, *Cinema 2: The Time-Image*, trans. Hugh Tomlinson and Robert Galeta, Minneapolis: University of Minnesota Press, 1989.

Deleuze, Gilles, *Bergsonism*, trans. Hugh Tomlinson and Barbara Habberjam, New York: Zone Books, 1991 [1966].

Duby, Georges (ed) *A History of Private Life* (five volumes), Harvard: Harvard University Press, 1988.

Eagleman, David, *Incognito: The Secret Lives of the Brain*, Melbourne: Text Publishing, 2011.

Edelman, Gerard, *Bright Air, Brilliant Fire: On the Matter of the Mind*, London: Penguin, 1994.

Eisenstein, Elisabeth, *The Printing Press as an Agent of Change: Communications and Cultural Transformations in Early-Modern Europe*, New York: Cambridge University Press, 1979.

Eliade, Mircea, *The Myth of the Eternal Return: Cosmos and History*, London: Arkana, 1989 [1954].

Ellis Smith, Robert, *Ben Webster's Web Site: Privacy and Curiosity from Plymouth Rock to the Internet*, Providence, RI: Privacy Journal, 2004.

Eriksen, Thomas Hylland, *Tyranny of the Moment: Fast and Slow Time in the Information Age*, London: Pluto, 2001.

Faulkner, William, *Requiem for a Nun*, New York: Vintage, 1975 [1950].

Fernández-Armesto, Felipe, *Ideas That Changed the World*, London: DK Publishing, 2004.

Fernyhough, Charles, *Pieces of Light: The New Science of Memory*, London: Profile Books, 2013.

Freeman, Charles, *The Greek Achievement: The Foundation of the Western World*, New York: Penguin, 2000.

Frugoni, Chiara, *Inventions of the Middle Ages*, trans. William McCuaig, London: Folio Society, 2010.

Garfield, Simon, *On the Map: Why the World Looks the Way It Does*, London: Profile Books, 2012.

Gergen, Kenneth G., 'Mobile Communication and the Transformation of the Democratic Process' in Katz, J. E. (ed) *Handbook of Mobile Communication Studies*, Cambridge, MA: MIT Press, 2008.

Giddens, Anthony, *The Consequences of Modernity*, Cambridge: Polity Press, 1990.

Goggin, Gerard and Crawford, Kate, 'Generation Disconnections: Youth Culture and Mobile Communication' in Ling, Rich and Campbell, Scott W. (eds) *Mobile Communication: Bringing Us Together and Tearing Us Apart*, New Brunswick: Transaction, 2011.

Goody, Jack, *The Domestication of the Savage Mind*, Oxford: Oxford University Press, 1977.

Gorman, Lyn and Mclean, David, *Media and Society into the Twenty-First Century: A Historical Introduction*, Second Edition, Chichester: Wiley-Blackwell, 2009.

Groys, Boris, 'Two Cures: Making Art Visible' in Merewether, Charles and Potts, John (eds) *After the Event: New Perspectives on Art History*, Manchester: Manchester University Press, 2010.

Guerlac, Suzanne, *Thinking in Time: An Introduction to Henri Bergson*, Ithaca: Cornell University Press, 2006.

Habermas, Jurgen, 'Modernity: An Incomplete Project' in Foster, Hal (ed) *The Anti-Aesthetic: Essays on Postmodern Culture*, Port Townsend: Bay Press, 1983.

Habermas, Jurgen, *The Structural Transformation of the Public Sphere: An Inquiry into a Category of Bourgeois Society*, Cambridge, MA: MIT Press, 1991.

Harvey, David, *The Condition of Postmodernity*, Oxford: Blackwell, 1989.

Havelock, Eric A., *Preface to Plato*, Oxford: Basil Blackwell, 1963.

Hawking, Stephen, *A Brief History of Time: From the Big Bang to Black Holes*, London: Bantam, 1988.

Hawking, Stephen (ed) *On the Shoulders of Giants: The Great Works of Physics and Astronomy*, London: Penguin, 2002.

Fredric Jameson, 'Postmodernism and Consumer Society' in Foster, Hal (ed) *The Anti-Aesthetic: Essays on Postmodern Culture*, Port Townsend: Bay Press, 1983.

Habuchi, Ichiyo, 'Accelerating Reflexivity' in Okabe, Ito and Matsuda, Misa (eds) *Personal, Portable, Pedestrian*, Cambridge, MA: MIT Press, 2005.

Jameson, Fredric, 'The End of Temporality', *Critical Inquiry*, Vol. 29 No. 4, 2003, www.jstor.org/stable/10.1086/377726

Kalantzis-Cope, Phillip and Gherab-Martin, Karim (eds), *Emerging Digital Spaces in Contemporary Society: Properties of Technology*, Basingstoke: Palgrave Macmillan, 2011.

Kittler, Friedrich, 'The Mechanized Philosopher', in Rickels, Laurence A. (ed) *Looking After Nietzsche*, Albany: State University of New York Press, 1990.

Lasch, Christopher, *The True and Only Heaven: Progress and Its Critics*, New York: W. W. Norton, 1991.

Lee, Heejin and Liebenau, Jonathan, 'Time and the Internet at the Turn of the Millennium', *Time and Society*, Vol. 9 No. 1, 2000, pp. 43–56.

Lefebvre, Henri, *The Production of Space*, trans. Donald Nicholson-Smith, Oxford: Blackwell, 1991 [1974].

Lefebvre, Henri, *Rhythmanalysis: Space, Time and Everyday Life*, trans. Stuart Elden and Gerard Moore, London: Continuum, 2004 [1992].

Levinson, Paul, *Digital McLuhan*, London: Routledge, 1999.

Levitin, Daniel J., *The Organized Mind: Thinking Straight in the Age of Information Overload*, New York: Viking, 2015.

Lévi-Strauss, Claude, *The Raw and the Cooked: Introduction to a Science of Mythology* Vol. 1, trans. John and Doreen Weightman, New York: Harper Colophon, 1975 [1964].

Levy, Pierre, 'Toward Superlanguage', *International Symposium on Electronic Art 94* Catalogue, Helsinki: University of Art & Design, 1994.

Levy, Pierre, *Collective Intelligence: Mankind's Emerging World in Cyberspace*, trans. Robert Bonomo, Plenum, New York, 1997.

Ling, Rich, *New Tech, New Times: How Mobile Communication Is Reshaping Social Cohesion*, Cambridge, MA: MIT Press, 2008.

Ling, Rich and Campbell, Scott W. (eds) *The Reconstruction of Space and Time: Mobile Communication Practices*, New Brunswick: Transaction, 2009.

Ling, Rich and Campbell, Scott W. (eds) *Mobile Communication: Bringing Us Together and Tearing Us Apart*, New Brunswick: Transaction, 2011.

Lippard, Lucy, *The Lure of the Local: Senses of Place in a Multicentred Society*, New York: New York Press, 1997.

Lloyd, Genevieve, *Being in Time: Selves and Narrators in Philosophy and Literature*, London: Routledge, 1993.

Lyotard, Jean-François, *The Postmodern Condition: A Report on Knowledge*, trans. Geoff Bennington and Brian Massumi, Minneapolis: University of Minnesota Press, 1984 [1979].

McGrath, John, *Loving Big Brother: Performance, Privacy and Surveillance Space*, London: Routledge, 2014.

McKay, Deirdre, 'On the Face of Facebook: Historical Images and Personhood in Filipino Social Networking', *History and Anthropology*, Vol. 21 No. 4, 2010, pp. 479–498.

McLuhan, Marshall (with Quentin Fiore) *The Medium Is the Massage*, Harmondsworth: Penguin, 1967.

McLuhan, Marshall, *Understanding Media*, London: Abacus, 1974 [1964].

McQuire, Scott, *Visions of Modernity: Representation, Memory, Time and Space in the Age of the Camera*, London: Sage, 1998.

Marinetti, F. T., *Selected Writings*, ed. R. W. Flint, New York: Farrar, Strauss and Giroux, 1972.

Martin, Adrian, 'Chopper: An Essay Around and About Shaun Gladwell,' in *Shaun Gladwell: Stereo Sequences* Catalogue, Melbourne: Australian Centre for the Moving Image, 2011.

Meikle, Graham and Young, Sherman, *Media Convergence: Networked Digital Media in Everyday Life*, Basingstoke: Palgrave Macmillan, 2012.

Merewether, Charles and Potts, John (eds) *After the Event: New Perspectives on Art History*, Manchester: Manchester University Press, 2010.

Meyrowitz, Joshua, *No Sense of Place: The Impact of Electronic Media on Social Behavior*, Oxford: Oxford University Press, 1985.

Milne, Esther, *Letters, Postcards, Email: Technologies of Presence*, New York: Routledge, 2010.

Miranda, Maria, 'Uncertain Spaces: Artists' Exploration of New Socialities in Mediated Public Space', *Scan Online Journal of Media Arts Culture*, Vol. 4 No. 3, December 2007 at http://scan.net.au.

Moores, Shaun, *Media, Place and Mobility*, Basingstoke: Palgrave Macmillan, 2012.

Morgan, Jessica and Muir, Gregor, *Time Zones: Recent Film and Video*, London: Tate Publishing, 2004.

Mumford, Lewis, *Technics and Civilisation*, London: Routledge, 1934.

Murphie, Andrew and Potts, John, *Culture and Technology*, Basingstoke: Palgrave Macmillan, 2003.

Niemeyer, Katharina (ed) *Media and Nostalgia: Yearning for the Past, Present and Future*, Basingstoke: Palgrave Macmillan, 2014.

Nietzsche, Friedrich, *Untimely Meditations*, trans. R. J. Hollingdale, Cambridge: Cambridge University Press, 1983 [1876].

Olalquiaga, Celeste, *Megalopolis: Contemporary Cultural Sensibilities*, Minneapolis: University Of Minnesota Press, 1992.
Ong, Walter J., *Orality and Literacy: The Technologizing of the Word*, London: Routledge, 1982.
Parika, Jussi, *What Is Media Archaeology?* Cambridge: Polity Press, 2012.
Parrinder, Patrick, *Shadows of the Future: H. G. Wells, Science Fiction and Prophecy*, Liverpool: Liverpool University Press, 1995.
Penrose, Roger, *The Emperor's New Mind: Concerning Computers, Minds, and the Laws of Physics*, London: Vintage, 1990.
Penrose, Roger, *Cycles of Time: An Extraordinary New View of the Universe*, London: Bodley Head, 2010.
Perkins, Maureen, *The Reform of Time: Magic and Modernity*, London: Pluto, 2001.
Petterson, Per, *I Curse the River of Time*, trans. Charlotte Barslund, Minneapolis: Graywolf Press, 2010.
Phelan, Peggy, *Unmarked: The Politics of Performance*, New York: Routledge, 1993.
Pickles, John, *A History of Spaces: Cartographic Reason, Mapping and the Geo-Coded World*, London: Routledge, 2004.
Pogue Harrison, Robert, *Gardens: An Essay on the Human Condition*, Chicago: University of Chicago Press, 2008.
Pohlmann, Ken C., *Principles of Digital Audio*, Indianapolis: Sams, 1992.
Potts, John, 'Who's Afraid of Technological Determinism: Another Look at Medium Theory', *Fibreculture Journal*, Issue 12, June 2008, http://twelve.fibreculturejournal.org/fcj-084-who's-afraid-of-technological-determinism-another-look-at-medium-theory/.
Potts, John (ed) *The Future of Writing*, Basingstoke: Palgrave Pivot, 2014.
Pound, Ezra, *ABC of Reading*, New York: New Directions, 1934.
Powell, Helen, *Stop the Clocks! Time and Narrative in Cinema*, London: I. B. Tauris, 2012.
Relph, Edward, *Place and Placelessness*, London: Pion Limited, 1976.
Richards, E. G., *Mapping Time: The Calendar and Its History*, Oxford: Oxford University Press, 1998.
Reynolds, Simon, *Retromania: Pop Culture's Addiction to Its Own Past*, New York: Faber and Faber, 2011.
Rose, Steven, *The Making of Memory*, London: Bantam Press, 1991.
Rossignol, Derrick, 'A New Invention Will Finally Prevent Cellphones from Ruining Concerts', *Music.Mic*, September 30, 2014. http://mic.com/articles/100048/this-technology-actually-keeps-people-from-ruining-concerts-with-cellphones. Accessed March 16, 2015.
Rush, Michael, *Video Art*, London: Thames & Hudson, 2003.
Rybczynnski, Witold, *Home: A Short History of an Idea*, London: Pocket Books, 2001.
Schafer, R. Murray, *The Tuning of the World*, New York: Alfred A. Knopf, 1977.
Scheer, Edward, *The Infinity Machine: Mike Parr's Performance Art 1971–2005*, Melbourne: Schwartz City, 2009.

Scheer, Edward, '"What If Somebody in New Zealand Wants to See It?" Performance Art's Cover Versions', in Merewether, Charles and Potts, John (eds) *After the Event: New Perspectives on Art History*, Manchester: Manchester University Press, 2010.

Schmitt, Carl, *The Nomos of the Earth*, trans. G. L. Ulmen, New York: Telos Press, 20013 [1950].

Seid, Steve, 'Introduction' in Anker, Steve, Gevitz, Kathy and Seid, Steve (eds) *Radical Light: Alternative Film and Video in the San Francisco Bay Area, 1945–2000*, Berkeley: University of California Press, 2010.

Seid, Steve, 'Thinking Inside the Box: Conceptual Video in the Seventies' in Anker, Steve, Gevitz, Kathy and Seid, Steve (eds) *Radical Light: Alternative Film and Video in the San Francisco Bay Area, 1945–2000*, Berkeley: University of California Press, 2010.

Silverstone, Roger, *Media and Morality: On the Rise of the Mediapolis*, London: Wiley, 2006.

Sinclair, Iain, *London Orbital*, London: Granta, 2002.

Solnit, Rebecca, *Infinite City: A San Francisco Atlas*, Berkeley: University of California Press, 2010.

Stravinsky, Igor, *Poetics of Music in the Form of Six Lessons*, Cambridge: Harvard University Press, 1970.

Taylor, J. C. (ed) *Futurism*, New York: Museum of Modern Art, 1961.

Thompson, E. P., 'Time, Work and Industrial Capitalism', *Past and Present*, Vol. 38, 1967, pp. 56–97.

Tsatsou, Panayiota, 'Reconceptualising "Time" and "Space" in the Era of Electronic Media and Communications', *PLATFORM: Journal of Media and Communication* Vol. 1, July 2009, pp. 11–32.

Tuan, Yi-Fu, *Space and Place: The Perspective of Experience*, Minneapolis: Minnesota Press, 1977.

Tunnicliffe, Wayne (ed) *John Kaldor Family Collection: Art Gallery of New South Wales*, Sydney: Art Gallery of NSW, 2011.

Turkle, Sherry, *Alone Together: Why We Expect More from Technology and Less from Each Other*, New York: Basic Books, 2011.

Varnelis, Kazys and Friedberg, Anne, 'Place: The Networking of Public Space' in Varnelis, Kazys (ed) *Networked Publics*, Cambridge, MA: MIT Press, 2008.

Virilio, Paul and Lotringer, Sylvere, *Pure War*, trans. Mark Polizotti, New York: Semiotext(e), 1983.

Virilio, Paul, *Speed and Politics*, trans. Mark Polizotti, New York: Semiotext(e), 1986 [1977].

Virilio, Paul, *The Virilio Reader*, Derian, James Der (ed) Malden: Blackwell, 1998.

von Franz, Marie-Louise, *Creation Myths*, Zurich: Spring Publications, 1972.

Wertheim, Margaret, *The Pearly Gates of Cyberspace: A History of Space from Dante to the Internet*, Sydney: Doubleday, 1999.

White, Hayden, *The Content of the Form: Narrative Discourse and Historical Representation*, Baltimore: Johns Hopkins University Press, 1987.

Index

Note: Locators followed by n refers to notes.

www.ingramcontent.com/pod-product-compliance
Lightning Source LLC
Chambersburg PA
CBHW060750150926
51501CB00004B/143

9781137494375